Young Crew

CW00762123

Also by John Mellor:

The Motor Cruising Manual
Boathandling Under Power
Sailing Can Be Simple
Cruising - Safe and Simple
The Sailing Cruiser Manual
A Small Boat Guide to the Rules of the Road
The Art of Pilotage
Logbook for Cruising under Sail
Cruising Skipper

and contributions to:

The Best of Sail Cruising
The Best of Sail Navigation

Young Crew

John Mellor

WATERLINE

Dedication

To Meredyth, Henry, Charlie and Rio
– my brother's young crew.

Acknowledgements

Acknowledgements are made to the following
for permission to reproduce illustrations.

Hydrographer of the Navy –
Chart of Runswick Bay (Page 119)

Fernhurst Books –
Illustrations from my work
'Logbook for Cruising Under Sail' (Appendix 2)

R & F Dearn –
The Shipping Forecast Chart (Page 148)

Published by Waterline Books
an imprint of Airlife Publishing Ltd
101 Longden Rd, Shrewsbury, England

© John Mellor 1993

ISBN 1 85310 373 X

A Sheerstrake production.

A CIP catalogue record of this book
is available from the British Library

Contents

You do not need reams of theoretical knowledge in order to be a useful crew aboard a yacht. Enthusiasm and a sound grasp of basic seamanship are the qualities a skipper most wants in a crew.

Introduction

In the old days, before fibreglass and sailing schools were invented, we learnt how to handle boats by sailing as crew with parents or older friends. This was a slower process than a crammed week at a sailing school, but in many respects gave a more solid understanding of the skills. It may have taken longer to learn how to helm a sailing boat, but on the way we picked up a great deal of basic seamanship such as dinghy handling, the observed effects of tide and weather, the practical use of knots, and a natural feel for being on and around boats. These are the things that cannot be absorbed during a concentrated sailing school course.

This book approaches the business of learning in the traditional manner, developing the skills required for good, plain seamanship - basic skills that are too often lacking in modern sailors. You will not find thirty-four different ways of gybing a spinnaker in here, but you should learn how to be the sort of thoroughly reliable and competent crew that an experienced skipper will be extremely pleased to have on board.

Part One deals with the experience and knowledge that you should pick up while pottering about in a dinghy - simple, basic seamanship; a general understanding of the effects of wind and tide; the skills required for the handling of small dinghies using various forms of propulsion. This should give you a firm base from which to move onto sailing aboard the family cruiser in Part Two, and hopefully help you to develop that natural feel for boats that only ever seems to come through an initial experience with the instability of dinghies.

A great deal of a yachtsman's time is spent messing about in tenders for one reason or another, and the competent handling of them is an essential attribute of a good crew.

Part 1 - Handling the Tender

Chapter 1
Basic Dinghy Work

Your first water-borne experience will hopefully be in the yacht's *TENDER*. This is a very small boat, generally called a *DINGHY*, which is used to get from the yacht to the shore, and is either carried on board or towed when at sea. When you become proficient at handling it you will be able to row ashore to explore or visit shops, collect water and stores for the boat and so on. You can have a great deal of fun in the dinghy as well as learning good basic seamanship, but you must always bear in mind that such a boat is small and unstable, and can be extremely dangerous if not handled properly. More sailors are drowned from dinghies each year than ever are lost at sea.

The reason I use the word 'hopefully' in the first paragraph is because the best way to initially learn about boats is unquestionably in a small, unstable dinghy. The reason is that such a boat will show you, immediately and unmistakably, the result of even the slightest thing you do to it. Move your weight too far out from the middle and it will promptly tip over, perhaps even capsize completely. Waggle the *RUDDER* (a piece of flat wood that steers it) and it will swerve instantly from side to side. The very instability of a small dinghy keeps your nerves tense, alert for the slightest miscalculation so that you can correct it rapidly, and keeps you on the look-out for approaching trouble so that you can forestall it. You will very quickly learn what actions on your part have what effect on the boat, and also the fact that you must at all times be prepared for any eventuality.

The same problems arise with larger boats of course, but unless experience with dinghies has taught you what subtle signs to look for, the boat just simply will not tell you as rapidly or as sensitively as a small dinghy will. A very large boat may produce such delays and distance between your actions and their results that as a beginner you may not even realise the results actually were caused by your actions.

So try to get as much experience as you can in the dinghy, in all conditions, and in friends' dinghies of different types as well, before going to sea in a cruising yacht. Any boat, even the biggest and most lumbering old cargo carrier, will speak to you through its movement, its sounds, its steering, the hiss of its bow-wave, the howl of its rigging, the clatter of its blocks, the shape of its sails, but it will not do so with the immediacy, the urgency and the clarity that a dinghy will.

Experience with the dinghy, however, will teach you how to read the messages of even the most inarticulate and truculent of big ships. Whether it is this truculence, or perhaps the affection they induce in sailors, that causes boats to be referred to as 'she' rather than 'it', I do not know. However, they are, and probably have been since the first one was ever built, so we will continue the tradition from here on.

Types of Dinghy

There are many different types of dinghy that you might come across, but by and large they can be divided into two main groups - rigid and inflatable. The first is solidly constructed of usually fibreglass or wood and cannot be dismantled. The second is flexibly built with tubes of a special type of strong and durable rubber. These tubes can be blown up with a pump to make the dinghy reasonably rigid for use, and deflated so that it can be rolled up and stowed away in a small space when not required. This type is similar in principle to plastic inflatable beach dinghies, but it is much stronger and more complex in design. Beach dinghies are most certainly not suitable for use as a yacht's tender.

You can see examples of these types of dinghy in Photos 1 & 2, and the names of the parts of a dinghy in Figure 1. These strange names (and you will find many more on board the yacht) may seem unnecessarily complex at first, but in truth they simplify things.

Photo 1 A variety of tenders lying on the beach - some fibreglass, some wood and some rubber. Some with engines, some with oars and some with sails. Details of all these types will be found in the first part of this book.

Photo 2 A rubber dinghy close-up. Note the blow-up seat and the built in *ROWLOCKS* (see later in this chapter) to take the oars when rowing. This can be easily lifted by one person, so it is not hard to imagine it blowing about on the water. Compare the dinghies in photos 1 and 2 and the photo at the beginning of the chapter.

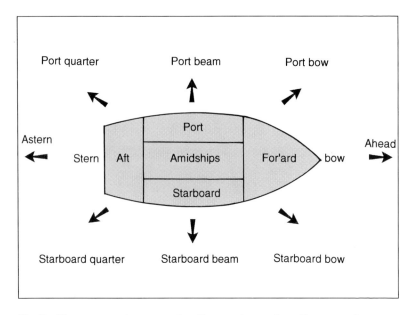

Fig 1a The correct names for the various directions and positions around a boat.

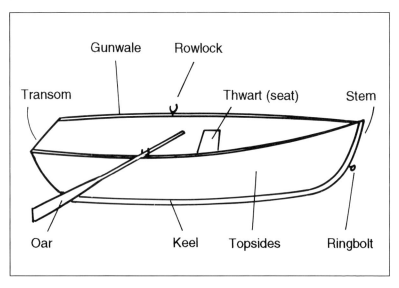

Fig 1b A typical small fibreglass tender, showing the proper names for the various parts of it.

GUNWALE, for example, is considerably more concise and accurate than 'the bit of boat at the top of each side'. If you make a point of always using the correct terms, even if it means constantly looking them up at first, they will soon become second nature. Diagrams showing the names of nautical things will be found throughout the book, and at the end is a comprehensive glossary of terms. (Words shown in capital italic letters eg. GUNWALE are the terms that you should try to remember now - a fuller list of sailing terms can be found on page 157)

Rigid dinghies, because of their shape and weight, are more difficult than inflatables to STOW (put away securely) on board and to haul up and down beaches, but they are easier to handle on the water as they sit more firmly in it and the KEEL (see Figure 1) guides them in a straight line as they move forward. Lightweight inflatables tend to bounce and blow around skittishly on top of the water and have no keel to keep them straight, so they can be difficult to row. They are, however, more stable and less likely to capsize than rigid dinghies. All dinghies are relatively unstable on the water so certain seamanlike precautions must be taken when handling them.

Safety Precautions

1 Always step directly into the middle of a dinghy - gently and steadily. Never jump in; and do not step onto the seats or the sides. The higher up your weight is, and the closer to the side it is, the more likely you are to capsize the dinghy. Once you are in, sit down immediately and keep still; too much movement will rock the boat and it could capsize. At all times the dinghy should be trimmed so that it sits level in the water: not leaning to one side, and not down at one end. Load passengers one at a time and spread them around in order to distribute the weight evenly. The dinghy should not be overloaded so that it sits too deeply in the water. The skipper of the yacht will advise on this as it requires experience to judge.

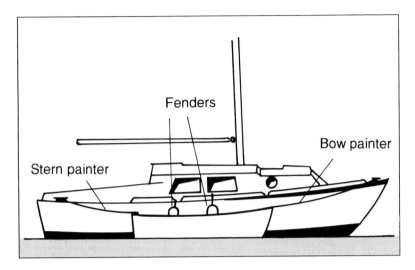

Fig 2 To hold the dinghy firmly alongside, the bow and stern painters must be secured fairly tightly. Making them as long as you can provides a little give to allow for the boat's movements.

2   Hold the dinghy very firmly while getting in or out so that it cannot drift away leaving you only half aboard. Try to step vertically downwards so that you do not push it away with your foot as you get in. If possible, tie it up securely by the *PAINTERS* (mooring ropes) at *BOW* (front) and *STERN* (back) so that it cannot move. This is particularly important when alongside the yacht out in the harbour, as you could get carried away by the current very rapidly if you fall in. See Figure 2. The correct knots to use are described in Chapter 5.

3   A *LIFEJACKET* or *BUOYANCY AID* should be worn at all times in a dinghy, as it will keep you afloat if you fall in (see Photos 3 and 4). A torch should be carried at night and waved about regularly so that other boats can see where you are. Do not shine it into anyone's eyes or it will take them nearly half an hour to get accustomed to seeing efficiently in the dark again! When your eyes are used to the dark (this is known as *NIGHT VISION*) you should not need the torch to

Photo 3 (Above left) This is a buoyancy aid: a simple buoyant waistcoat that will help you stay afloat in the water. It has a minimum of buoyancy and no buoyant collar, so will not keep an unconscious person afloat with their face clear of the water. For most small boat work in reasonably sheltered waters it is a useful aid that is not too awkward to wear.

Photo 4 (Above right) This is a proper lifejacket, albeit a simple one. When inflated it opens out across the chest and the collar also inflates to hold the head above water. It is then a lot more effective than the buoyancy aid, but more of an encumbrance.

see where you are going. Always tell the skipper where you are going to in the dinghy and when you expect to return.

Photo 5 A traditional wooden dinghy alongside a small yacht. Although a bit tatty, it is well equipped as a tender, containing a pair of oars plus spare, anchor and warp, long and short painters at both bow and stern, large and small bailers, outboard motor with spare parts, tools, spare fuel and a funnel. Every one of these items is tied securely to the dinghy in case of capsize or swamping.

4   The following equipment should be carried in the dinghy at all times: bow and stern painters; *ANCHOR* and *WARP* (the anchor line), with the end of the warp secured to a strong point in the bows of the dinghy; a pair of oars plus a spare; a pair of *ROWLOCKS* plus a spare; and a bailer (to bail out any water). An anchor is a specially-shaped lump of metal that can be lowered to the seabed on the end of a line (anchor warp), where it will dig in and stop the boat from drifting. A rowlock is a U-shaped metal fitting in which the oar sits when rowing - see Figure 1 and Chapter 2. All these items should be tied to the dinghy, so that they will not be lost in the event of a capsize. Inflatables should also carry a pump and a repair kit, in case of damage or deflation. See Photo 5.

5   Finally, always carry a sharp knife whenever you are in or around boats, preferably a clasp knife incorporating some device for undoing the pins of *SHACKLES*, and small enough to fit in your pocket. Shackles are small U-shaped metal objects that close with a pin (generally screwed through one end of the U and into the other); they are used all over boats for joining things together and you will find yourself constantly doing them up and undoing them. See Figure 3. The skipper will advise you on the most suitable equipment for the shackles on his boat. Keep the knife sharp on an *OILSTONE*; it may save your life one day if you find yourself trapped by a rope that needs to be cut quickly.

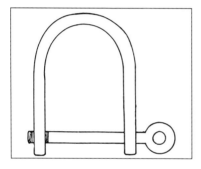

Fig 3 A common type of shackle.

Getting out of Trouble

This is perhaps rather early in the book to be discussing such a topic, but there are some aspects that can be considered now. The first is the importance of following the advice above, in order to avoid getting into trouble in the first place. It is very easy when on the water for a small problem to develop rapidly into a catastrophe if it is not dealt with immediately and decisively. If a difficulty cannot be resolved right away then you must prevent it from worsening while you think out how to deal with it, or call for help as the case may be. One of the best ways of preventing trouble worsening is simply to anchor the dinghy so that it cannot be carried away into danger. Keep the anchor and warp tidy and ready for use at all times so that this can be done quickly and without tangles. See Chapter 5.

If you do get in trouble and cannot rescue yourself you should call for help by raising your arms repeatedly up to the horizontal and back down again. This is a recognised distress signal (see Appendix 3). At night you should flash the letters S O S with the torch in Morse code. This consists of three short flashes, three long flashes and three more short flashes. Pause a while then repeat. Aim the torch towards where you think there might be people watching: a nearby boat showing cabin lights; a pub or house onshore. NEVER leave the dinghy to swim ashore unless it actually sinks beneath you. You stand a much better chance of survival if you stay in the boat and make distress signals.

Perhaps one of the most serious problems you can encounter in the dinghy is someone falling overboard. If you have the strength to get them back on board then pull them up over the stern, where the dinghy will be most stable; if you try to pull them over the side you will capsize. If you do not have the strength to pull them aboard, then tie them to the stern of the dinghy with the painter and hold their head clear of the water while signalling for help. If necessary, anchor the dinghy; but if they can hold on to the *TRANSOM* (see Figure 1) then consider rowing for the shore, dragging them behind.

If someone is in the water for any length of time they will become extremely cold, and may even need hospital treatment to prevent death from what is called *HYPOTHERMIA*. Get advice immediately from a yacht skipper, harbourmaster or policeman, meanwhile wrapping the victim up as warmly as possible and keeping him out of the chilling effect of the wind. Have someone cuddle up to them to generate heat.

Taking Care of the Dinghy

Dinghies are expensive items and so are the various bits of equipment carried in them - especially *OUTBOARD MOTORS*. The latter in particular are vulnerable to theft when a dinghy is tied up ashore or left on a beach. If you can find somewhere safe to leave outboard, oars and rowlocks (ask the Harbour Master) then remove them from the boat when you get ashore. If you cannot, then try to leave the dinghy somewhere busy where a thief is likely to be noticed. Wedge or tie the oars in firmly so the time that would be needed to remove them will deter a casual thief. Outboards can be chained to a strong point at the stern and padlocked.

If the condition of the dinghy is allowed to deteriorate too much it could become dangerous or uncomfortable to use. Do not drag it along beaches or up slipways, unless it is a rigid dinghy with a metal band on the keel; then hold it upright so that it slides along this metal *KEELBAND*. There should be *FENDERING* around the gunwale to protect it from bashing against jetties etc, and this needs to be kept firmly secured.

A good seaman takes a pride in his boat, so keep her clean and tidy and make sure everything works properly - *SHIPSHAPE* is the expression. Stow anchors and warps out of the way so you do not trip over them, and if they get muddy clean them thoroughly before bringing them on board, or the mud will get everywhere. The same goes for your boots. Sand is very abrasive and will scratch paint and damage rubber if it is left lying about in the boat. Tip the dinghy on its side and wash the sand out by throwing water round the inside using the bailer, or find a hosepipe. Remember that 'a ship is known by her boats', so don't let the ship down by going ashore in a scruffy boat.

Chapter 2
Rowing and Sculling

These are the two commonest, and most useful ways of moving a dinghy about using manpower alone. The first is very well known and is simple to do, and is probably the least tiring of all the techniques (if performed correctly). It can, however, be awkward when manoeuvring in confined spaces, in which situation *SCULLING* is a great deal more convenient. Sculling is difficult until one grasps the knack of it, is tiring on the arms and is much less efficient than rowing, but is an extremely useful skill to acquire, as, besides its convenience when manoeuvring, it has the great benefit in emergency of needing only one oar and no rowlocks.

Rowing the Dinghy

Dinghies are generally rowed most of the time, this being a cheap, simple and reliable method of propulsion. A good rowing technique is most important as not only does it make the work easier, but it also enables you to row faster and for longer, which can be essential to your safety in strong winds and currents (see Chapter 4).

There is nothing difficult about the basic principle of rowing. You sit on the centre THWART (seat) facing AFT, as it is easier to pull back on an oar than to push forward on it. You place the oars in the rowlocks (which fit into the gunwales), lean right forward holding the handles, dip the blades into the water, then pull back on the oars together. The blades should be more or less vertical in the water for maximum pull, although if you tilt the bottoms slightly aft you will find it easier to pull them out of the water in readiness for leaning forward again to make the next stroke.

Do not tilt the blades too far or the oars will tend to 'dive' down into the water rather than pull straight along. If you tilt the tops aft then the blades will tend to jump out of the water, much as a skimming stone does. This is known as 'catching a crab' and the sudden loss of water pressure against the blade can cause you to fall backwards off the thwart, to the amusement of anyone watching. At the beginning of each stroke you should keep your arms straight and pull on the oar by leaning back with your trunk, bracing your feet firmly in order to transmit the force to the boat through your legs rather than your bottom.

The blades should not be dipped any deeper than necessary or you will find they 'jam' in the water and you cannot get them out again! Short, quick strokes are best for getting started, then gradually ease into a steady rhythm of long strokes, leaning as far forward as you can to begin each one then leaning back as far as possible to continue it. Rubber dinghies, however, are usually best

Photo 6 A young boy rowing a solid, fibreglass dinghy in choppy waters. He should be wearing a lifejacket. Note the level trim of the boat in the water.

Photo 7 Rowing a small rubber dinghy in a sheltered creek. The rope round the edge is to hold onto if you are in the water. Note the position further forward for the rowlocks to fit into; this enables you to row whilst sitting in the bow and thus keep the boat trimmed level if there are two of you on board.

rowed with short strokes all the time as they are too light to maintain the momentum required between long strokes. See Photos 6 and 7.

As you pull the blades clear of the water in readiness for the next stroke you should bend your hands back at the wrists so that the blades twist to become level with the water. Hold them like this while you lean forward and they will create less resistance to the wind. Twist them back into the near vertical position just before entering the water for the next stroke. This may sound like fussy specialist rower's technique, but if you are rowing into a strong wind it will make a tremendous difference to your progress. This process is known as *FEATHERING* the blades, and if you do not believe the effect of it then try rowing into a strong wind without doing so! You will soon learn.

During all this complicated manoeuvring and concentration you must also keep a good look-out all around for approaching boats, rocks, jetties and so on. You will have to glance frequently over your shoulder to see where you are going, and at the same time pull evenly on both oars so that you travel in a straight line. This is not always as easy as it sounds as a number of things can cause you to pull more heavily on one side than the other. The most likely are: one arm stronger than the other; a side-wind tending to swing the dinghy; one oar immersed deeper, or further out from the boat than the other; the boat listed to one side; the boat trimmed down by the bow. Practice helps, naturally, but it is important to understand the things that can make rowing difficult. See Figure 4.

There will of course be times when you want to turn and manoeuvre the dinghy other than in a straight line. Turning is done simply by slowing down or stopping the oar on the side towards which you wish to turn. The opposite oar will then swing the boat. You can turn very sharply - on the spot if necessary - by pushing backwards on the oar inside the turn and normally on the one outside. This takes some practice, but it will enable you to manoeuvre the dinghy into and out of the tightest spots. You can even row the dinghy backwards by pushing on both oars instead of pulling. It is

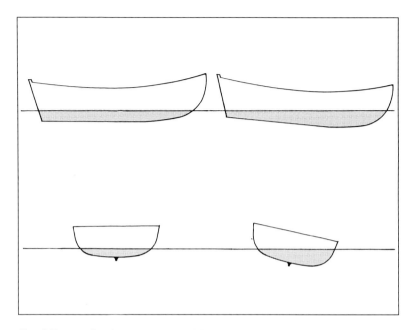

Fig 4 For a dinghy to row well the aft end of its keel must be deeper in the water than the forward end. If the boat then attempts to swing off course the pressure of water against this deeper end pushes it back straight. If you trim a dinghy down by the head(bow), however, the pressure of water against the deeper forward end will accentuate any slight swing and it will be almost impossible to control. If the dinghy lists to one side, the underwater shape changes and this will make it turn off course all the time.

extremely useful to have this sort of versatile control over the dinghy, and you should practise whenever you can.

However good your rowing technique, you will not be able to row properly if the dinghy is not organised efficiently for it. The rowlocks must be firm so that they do not move when you pull against them, or you will find the oars jumping out all the time. Rowlocks often have one horn longer than the other: set the longer horn for'ard so that the oar bears against it when you pull. The oars must be of a length that, with the blades just dipped in the water, the handles meet your hands at a comfortable height and do not overlap at the ends; generally half as long again as the beam of the dinghy. There must be something solid for you to brace your feet against as you pull back on the oars.

The oars should never be left trailing in the water hanging from the rowlocks - however careful you are, sooner or later you will lose one. *SHIP* them when they are not needed, by swinging the handles forward then sliding the oars forward in the rowlocks so that the handles lie in the bottom of the dinghy (or on the for'ard thwart) and the tops of the blades rest in the rowlocks. If you go alongside a boat or jetty the oars must be lifted right out of the rowlocks and the rowlocks removed from the gunwales, and the whole lot laid in the boat, otherwise they will scratch or jam against the jetty or yacht. A good way of stowing them securely is with each rowlock and its safety line wrapped around its respective oar (see Photo 8).

Photo 8 A traditional wooden dinghy. Note the thick rope fender around the gunwale, the seating for passengers round the stern, the two rowing positions (middle seat and forward seat), the notch in the transom for the sculling oar. See how the lines holding the rowlocks to the boat are wrapped around the oars to secure them to the boat.

Sculling the Dinghy

Sculling a dinghy with one oar over the stern is a very useful skill as it enables you to manoeuvre and control the boat much more efficiently in tight spaces than does rowing. It is not hard to do, but it is very much a knack, like riding a bicycle or swimming. All three can be explained precisely, but when you come to carry out the instructions you either fall off, sink, or simply look silly with the oar bouncing up and down off the water and the boat going nowhere. Practice with any of these exercises does not produce a gradual improvement; you merely flounder about uselessly for a while, then suddenly get the hang of it. After that it is easy.

The principle of sculling is to rest the oar in a U-shaped notch in the transom, then pull the oar from side to side so that on each stroke the blade slides across at such an angle as to push the water backwards while pressing on the boat and thus pushing it forwards. Start with the oar just above the water to one side of the stern as you can see in Figure 5, then pull it down into the water and across to the centre of the transom by pushing up and over on the handle. Keep the blade at such an angle that it presses against the water and tries to push it backwards away from the boat; and move the handle at a fairly shallow angle so that the oar presses against the side of the rowlock. At the middle point twist your wrist so that the blade is horizontal, then swing it in an arc (with the thin edge leading all the time) until it is just below the surface on the other swing. See Figures 6 and 7. Then twist it back to the angle with the water that you started with and pull it down and across to the middle again in the same manner as the first down stroke. See Figure 8.

The stroke down into the middle of the boat does the driving, while the upward stroke simply gets the oar up into position for the next driving stroke from the other side. Initially the oar will tend to jump out of both rowlock and water, but with practice you will get a feel for just the right angle to exert the strokes, then you will find your hand swinging back and forth in a rhythmical figure-of-eight motion as you can see in Figure 9. If you practise

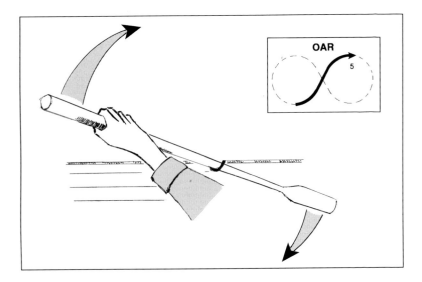

Fig 5 The arrow on the right shows the movement of the hand in this first part of the stroke. Note the position of the hand and fingers.

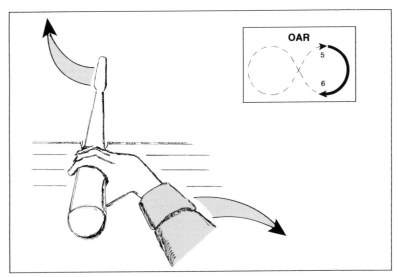

Fig 6 This is the mid-position with the hand up and the blade flat as shown. The hand movement is shown by the right hand arrow.

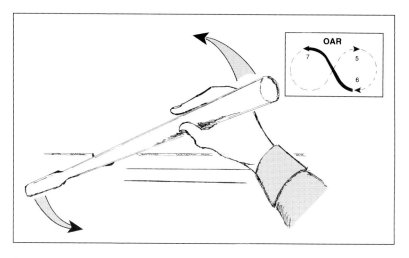

Fig 7 Having twisted the hand to the right and down, the blade is now close to the surface and pointing upwards. Note the position of the hand under the oar. Follow the solid right-hand arrow for the next move.

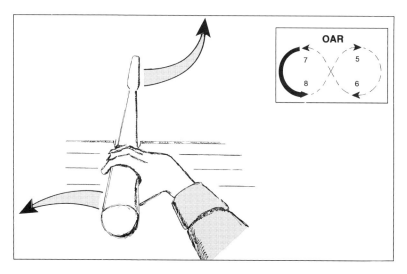

Fig 8 With the oar once again in the middle, but travelling the other way, this final part of the stroke brings the hand over to the left and down again ready to begin the next stroke as in figure 5.

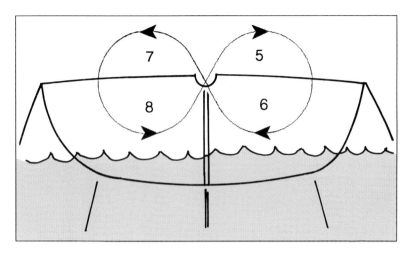

Fig 9 This shows the figure of eight movement that your hand should make as it works the oar through the various actions depicted in figures 5 to 8.

Photo 9 A young girl skilfully sculling the dinghy shown in photo 8. As this dinghy is heavy for its size, she sculls standing up and using both hands to get maximum power into the strokes. A smaller lightweight dinghy can be sculled single-handed while sitting down.

first with the dinghy firmly tied up by bow and stern so that it cannot move, you will find things a lot easier as the oar will be much less inclined to jump out, due to the boat not moving away from it all the time. You should be able to feel the pressure exerted by the oar on the driving stroke if you have the blade and the pull at the correct angle.

Sculling is best done standing up in a large dinghy (ten feet or so) as more power can be exerted on the oar. Small dinghies are safer sculled from a sitting position due to their instability. Holding the oar with two hands makes it less tiring but the rhythm is harder to maintain smoothly. Generally, small dinghies respond best to rapid, short strokes, and heavier dinghies to longer, steadier ones. Try all ways and see what suits you and your dinghy best. After this, practice will make perfect. The dinghy can be steered by exerting more pressure to the stroke on one side than the other, which is done by increasing the angle of the blade with the water. If the blade is pulled down with no angle (with the thin edge leading), then no drive will be produced. See Photo 9.

Sculling can also be done over the bow, which is rather easier both to do and to describe. It is more or less like paddling alternately on each side but without lifting the oar from the water. Kneel right forward and make a short paddling stroke down one side of the dinghy close to the bow. Then, instead of lifting the oar from the water, simply twist the blade so that it points forward and slide it forward through the water to the bow. Swing the blade round the bow, at the same time turning it so that it travels point first all the time. When it crosses the bow it will be in just the right position to make a paddling stroke on the other side. Keep doing this until you get to where you are going. The dinghy can be turned by repeated paddling strokes on the same side: as the oar slides forward to the bow, instead of swinging it round the bow and paddling down the other side, simply twist the blade flat and paddle back down the same side. With practice you can do this all ahead of the bow, using very short strokes that do not pass back along the boat. This is rarely-seen but a useful technique.

Chapter 3

Motoring and Sailing

There can often be times when the yacht is moored a long
way out from the shore and, particularly in strong currents
and winds (see Chapter 4), it can be hard work, or even
dangerous to try and row or scull ashore. An outboard
motor can then be used to propel the dinghy, or perhaps
even a small mast and sail. Such things will also enable you
to take the dinghy exploring up to the heads of creeks and
suchlike, journeys that would likely take all day with oars
alone.

The operation of outboard motors or of sails is rather
more complex than using oars, and also in certain respects
potentially more dangerous. Doubtless the yacht's skipper
will both advise you on the sense of using them in the
prevailing conditions, and also instruct you in their use. The
following information should, however, be helpful to you.

Outboard Motors

These are very small engines that can be clamped to the
stern of a dinghy. See Photo 10 and Figure 10. They are
generally fairly simple to operate, but with a sharp metal
PROPELLER whirling round just under the water and an
explosive fuel they can be highly dangerous and must be
handled with great care. They must never be run
anywhere near people in the water, due to the sharp
propeller blades, and should also be kept well clear of
ropes, weed, plastic bags and so on that could tangle in
the blades. Because of the risk of explosion from the fuel,
they should never be refuelled while running, or in the
presence of naked lights such as cigarettes. Care should
also be taken when approaching beaches or rocks, to
avoid hitting them with the propeller when the water
shallows.

When an outboard motor is stopped it can be tilted up
to keep the propeller clear of the water. This should be
done on mooring up and on approaching a shallow beach.
Always shut off the fuel and the breather in the

Photo 10 An outboard motor clamped to the transom of a dinghy and tilted clear of the water. This protects the propeller from the bottom and from passing debris, and also prevents it collecting weed. When not required it can be unclamped and stowed in the bottom of the boat. Note that it is fitted to one side of the sculling notch so that the latter can still be used.

filler cap before tilting, or petrol will pour out everywhere,(See next section). Starting and operating methods vary with the type of outboard, so consult the skipper or the manual. An outboard should be secured to the dinghy with a safety line at all times in case it falls over the side. The clamps that hold it to the transom must be done up as tightly as possible; if they slip while the engine is running it could jump right off the boat and cause serious damage to either the boat or its occupants.

Handling a dinghy with an outboard is not difficult but a few hints and tips will help. The first is to do your initial practising in calm, sheltered water a long way from expensive moored yachts. If you lose control while roaring along under outboard in the middle of closely-

31

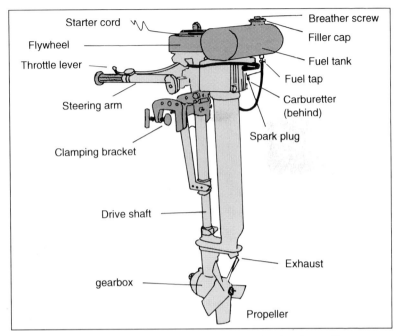

Fig 10 This shows the basic structure of a typical small outboard motor. The parts that concern your operation of it are labelled.

packed yacht moorings you can cause an awful lot of damage before getting things back under control. Bear in mind that a lightweight dinghy will be extremely responsive to the power of an outboard, so avoid sudden acceleration and turning or you could quite easily capsize, or throw yourself out. If the latter happens, the dinghy will career off all by itself and doubtless cause untold damage and trouble somewhere.

People often get into a muddle simply by sitting in the wrong place in the dinghy, such that the outboard steering arm is jammed into the small of the back, the boat is trimmed with the bow stuck up in the air, and so on. You must sit so that you can easily steer and operate the controls, and also so that the dinghy trims properly and thus will steer and manoeuvre predictably. See Chapters 1 and 2. If you are alone in the dinghy this

Photo 11 This small rubber dinghy is somewhat overloaded and on a windy night with waves in the harbour could be dangerously unstable. The crowding does tend to cramp the helmsman into an awkward steering position, which would contribute to the risk. Note the life-jackets on the youngsters, with the collars inflated round their necks.

invariably means sitting well forward so that you cannot reach the outboard! A simple extension steering arm will make life very much easier on a long run, although you will need to move aft to control the motor when coming alongside. See Photos 11 and 12.

When you are more experienced and confident you should experiment with steering by simply altering the trim of the dinghy. If you heel the dinghy to starboard, the alteration in underwater hull shape will cause it to turn to port, and vice versa. With practice this system will be found most effective, although it does need a dinghy that steers well in a straight line to begin with (ie one having a long keel that is deepest aft). See the comments in Chapter 1 about the effect of trim on steering.

Photo 12 This outboard motor's steering arm has an extension rod fixed into it, enabling the helmsman to sit forward and keep the dinghy trimmed properly if he is on his own.

Outboard Adjustment and Trouble-shooting

Small outboard motors are generally simple devices having few controls. A clear understanding of these controls will, however, make their operation most efficient and reliable. In principle the motor consists of two parts - the engine and the propeller. These are usually connected by means of a *CLUTCH*, which enables you to disconnect propeller from engine when you want to stop. Some very small outboards have no clutch, and roar off ahead the moment they start. To stop the dinghy this engine must be shut off completely. Some engines also have a *GEARBOX* which enables the propeller to turn the opposite way and so pull the dinghy backwards.

The *THROTTLE* controls the speed of the engine. The *CHOKE* allows extra fuel into the engine to make it start

34

more easily, and is closed when the engine has warmed up. The fuel is kept in a tank which has a tap to shut it off when the engine is not running. In the filler cap is a vent screw which allows air into the tank to replace the fuel as it is used up. This must also be kept closed when not required, in order to prevent the dangerous fuel leaking out and possibly causing a fire or explosion. The engine may be stopped by a special stop lever, or by closing the throttle, or by simply shutting off the fuel tap and letting it run out of fuel. There should be some means of adjusting the angle that the outboard shaft makes with the transom, so as to ensure that the propeller is pushing the dinghy directly forward and not up into the air or down into the water.

You should familiarise yourself thoroughly with the workings of your outboard, and learn how to use the simple spare parts that should be carried in the dinghy. There are not many things to go wrong with a small outboard, but they are much easier to fix if you know what they are and how they work. In principle it works by sucking in both air and fuel, then mixing them together and igniting the mixture with a spark from the spark plug. The simple things most likely to cause trouble are lack of fuel or lack of spark, and both can be tested quite simply, and nearly always mended also simply. Ask your skipper to show you how. The fuel used in nearly all small outboards has special oil mixed in it to lubricate the engine. This is called *TWO- STROKE* mixture, and if the motor is old and worn the oil can often settle on the spark plug and prevent a spark. A simple clean will thus cure the problem.

Sailing Dinghies

Large yachts often carry these as tenders and they can be a lot of fun to potter around creeks and rivers in. See Photos 13 and 14. Sailing dinghies, however, are even more unstable than rowing dinghies as the pressure of wind on their sails can literally blow them over. Generally you need to sit on the side opposite the sail, so that your weight balances its tendency to capsize the boat. See Figures 11 and 12.

Photo 13 A couple of small sailing dinghies that could be used as tenders for yachts. They are built of plywood and are light in weight, and can be sailed, rowed and motored easily and efficiently. They are a class known as Mirror Dinghies, and many clubs race them, which is a marvellous way of improving your sailing skills.

It is not difficult to learn to sail a small dinghy by trial and error, but a few hints here will help you. In principle, for the sail to drive the boat it must be pulled in by the *SHEET* (a simple control line) until it stops flapping. The wind is then blowing smoothly around the sail and producing energy that drives the boat. To slow down and stop you simply let the sheet out until the sail is flapping, whereupon it will cease to generate energy. The boat is steered by turning the rudder with the *TILLER*, the tiller being pushed in the direction you want the stern to swing. The bow will go the other way.

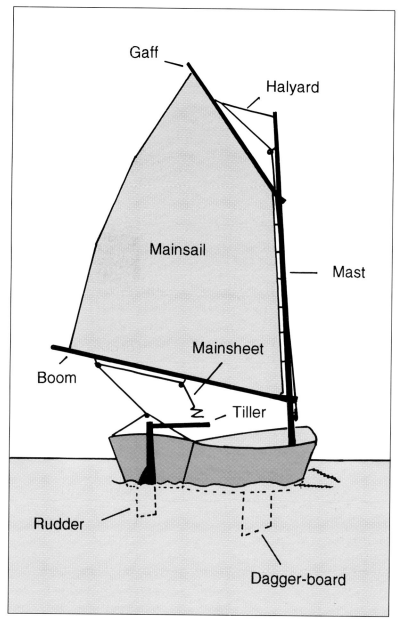

Fig 11 The proper names of the parts of a small sailing dinghy.

Photo 14 This is a different sort of sailing dinghy altogether, being rubber and specifically designed for use as a yacht's tender and also a liferaft. As with the Mirror Dinghy it can be rowed, sailed or motored, but it can also be deflated and stowed on board in a small space on deck or below. Unless a yacht is very large, a Mirror Dinghy would have to be towed.

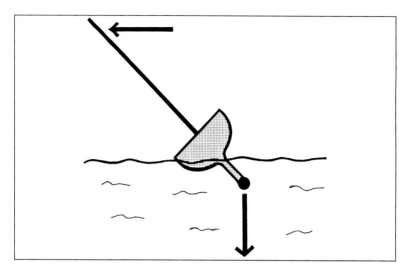

Fig 12 As the wind blows against the sail so it tends to heel the boat over. The picture shows how a yacht withstands this force: a heavy weight is fixed to the underneath of the keel and as the yacht heels over so this weight is lifted. The yacht is prevented from capsizing by the force of the weight trying to fall down again, as you can see from the arrows. A small dinghy does not have this weight so it needs the crew to lean out over the side opposite the sail so as to hold the boat upright.

The main problem with sailing is that you cannot sail straight towards the wind as the sail then simply flaps, however hard you haul it in with the sheet. If you need to go to *WINDWARD* (towards the wind) you must zig-zag back and forth with the sail pulled in tight, sailing as close to the wind as you can on each TACK (zig or zag). You must continually swing the boat very gently towards the wind until the sail begins to flap, then bear away from the wind until it fills again, so that your track is very slightly S-shaped. This enables you to be always as close as possible to the wind, and thus sailing most efficiently to windward.

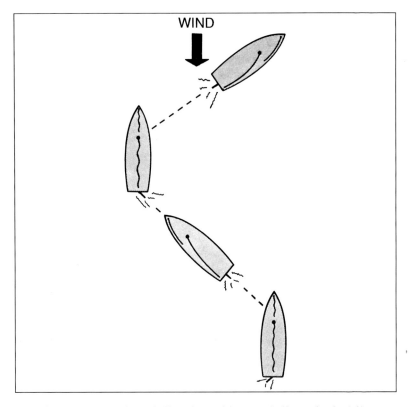

Fig 13 The dinghy is pointing head towards the wind at the bottom and you can see the sails flapping. As it turns away from the wind so the sail fills when about 45 degrees from the wind direction and the boat can sail along. To make ground directly to windward you must every now and then swing the boat through the wind and sail off on the 'other' tack as shown.

At the end of a tack (when you reach shore or a moored boat perhaps) you must push the tiller to *LEEWARD* (away from the wind) so that the boat swings towards the wind. Let her swing right round until the wind is on the opposite bow, move yourself across the boat to the other side, then continue sailing to windward as before, but with the wind on the other side of the dinghy. Figure 13 should make this process - called *TACKING* - clear.

When you sail directly away from the wind the sail is trimmed by letting it right out until it lies across the boat at right angles. In this position, with the wind blowing from astern, there will be no tendency for the boat to heel over so you must sit in the middle somewhere to keep her balanced. In between these two extremes - tacking into the wind and running before the wind - you should trim the sail with the sheet until it just stops flapping and no tighter. Sit where your weight will keep the boat upright, and also where you can control the tiller and sail comfortably as well as keep a good look-out. All this sounds highly complex when written down, but you will find as you practise the manoeuvres that it gradually becomes clearer. See Figure 14.

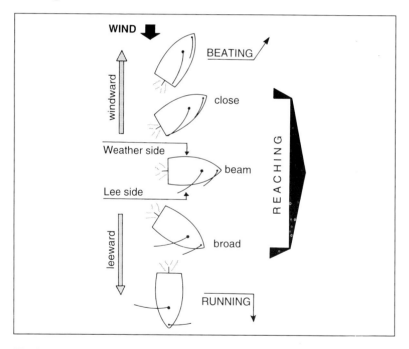

Fig 14 You can see here how the sails should be set for beating (to windward), reaching (across the wind) and running (away from the wind). Note the terminology with respect to the wind, and also the three types of reaching - close reach, beam reach and broad reach.

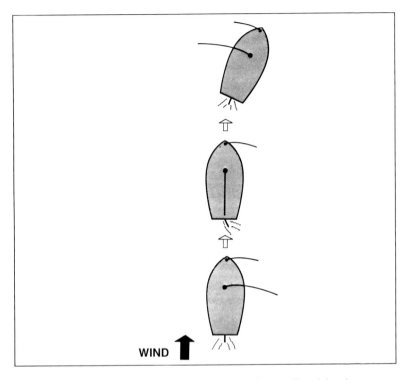

Fig 15 You can see the gybing process here, the black arrow showing the wind. If you have a jib as well as a mainsail this must change sides too and be hauled in with the other sheet after the mainsail has gybed.

If you are running with the wind right astern and the sail let fully out and you want to turn towards the wind you have simply to push the tiller to leeward and pull in the sail until it trims. If, however, you want to turn the other way, so that the wind swings across your stern, you will need to shift the complete sail from one side of the boat to the other. To do this without the wind getting behind it and slamming the sail across suddenly, the force of which could cause you to capsize, you should haul the sail tight into the middle of the boat first, then alter course. When the wind flips into the back of the sail you can let the sail out steadily on the new side until it is properly trimmed. See Figure 15. This is called *GYBING.*

When you get close to shore or in a tricky place you should lower the sail and row or scull to your destination. You will, however, find often that simply waggling the rudder from side to side will propel you slowly forwards. Just before grounding on a beach you should haul up the CENTREBOARD or DAGGER-BOARD so that it does not hit the bottom, and also haul up the rudder blade if it lifts, or remove the rudder and put it inside the boat. Have an oar handy to paddle or scull the last few yards.

Probably the most difficult thing newcomers to sailing find is determining precisely where the wind is coming from. With practice you will be able to feel it quite easily on your face or the back of your neck and gauge the direction with little trouble. In the meantime if you are uncertain, simply let the sheet run out until the sail is flapping completely, then think of the sail as a flag on a flagpole, blowing away from the mast in the direction the wind is going. Haul it in until it just stops flapping and the sail is now using the wind to drive the boat. With practice you will soon get the hang of constantly easing the sail out until it begins to flap, then hauling it in again to just put it to sleep. This keeps the sail working at its most efficient, and also enables you to keep track of where the wind is blowing from, as it is likely to vary slightly from time to time. Remember, however, that when sailing 'to windward' you should pull the sail in tight and leave it there, adjusting for changing wind direction by steering the boat so as to sail as close to the wind as you can without the sail flapping.

If the wind blows strongly you may find the sail too much to handle, and the boat feeling unstable to the point of possibly capsizing. If the sail is equipped with a REEFING system then you can reef it (make it smaller) so that it tends to heel the boat over less. In Figure 16 you can see the two reefing systems that you are most likely to come across. If the sail cannot be reefed then you must lower it and row back to the yacht or shore.

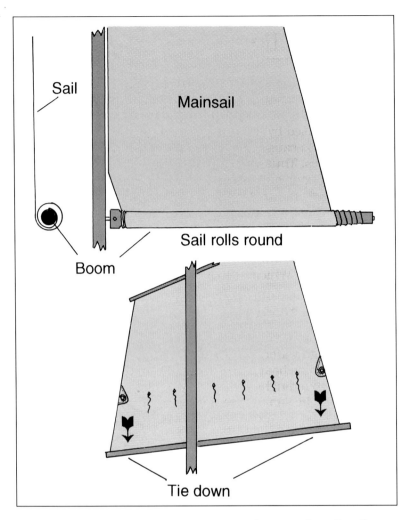

Sail

Mainsail

Sail rolls round

Boom

Tie down

Fig 16 There are two basic ways of reefing a sail - rolling it around the boom or tying part of it down to the boom. Proper equipment is needed for doing either of these on a yacht, but the small sail on a dinghy can be simply rolled round the boom by hand or tied down with short lengths of line if the sail has the system shown in the bottom picture. It is important that the strain is taken on the two eyes in the bottom system, the short lines in between being used just to tie up the surplus sail.

Chapter 4
Wind and Tide

The stronger the wind blows the bigger the waves will get, and the more difficult, and eventually dangerous it will be to go out in the dinghy. *TIDES*, which are movements of water in the sea caused by the gravitational pull of the sun and moon, can also cause dangerous conditions in certain circumstances. Thus you should begin your dinghy work in simple conditions of light winds and weak tides, then only gradually work up to harder conditions as you gain both experience and confidence; your skipper should guide you in this. In the meantime do not think it is clever to go out in bad weather: there is an old nautical saying that any fool can go out to sea, but it takes a seaman to know when to stay in harbour. Which means .... precisely what it says!

Wind and Waves

Even in quiet weather, however, you can experience what it is like in rough water. If some ignorant oaf roars past you at speed in a motor-boat, the *WASH* kicked up will be like rough seas to a dinghy like yours - it could capsize or swamp you. The same problem can occur with the wash of a ship passing up a channel. See Photo 15.

Photo 15 Note the waves (called wash or wake) left behind this ship as it proceeds up river. A dinghy could easily be swamped or capsized by such waves, even on a calm day. As they reach the shallows at the side of the river they could get bigger and break. See Figure 18.

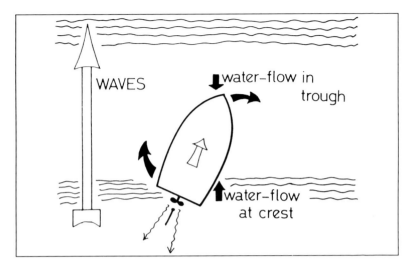

Fig 17 The water in a wave flows round the wave, moving forward at the crest and backwards in the trough. If the stern is picked up by a large wave, this flow will push it forward; at the same time the bow in the trough will be pushed back and the boat will tend to be twisted off course.

If these waves hit you on the side of the dinghy they could lift up the side and roll you right over, or at least crash over the side and fill you up. If they hit you on the flat transom that most dinghies have they could also swamp you, or they could lift up the stern and swing it round till you are *BEAM-ON* (sideways) to the next wave, which could then capsize you. This is known as *BROACHING*, and you can see what happens in Figure 17. The best way to ride these waves is to steer almost straight at them so that the bow rides up and over them at a slight angle. The sharp bow will push the wave aside and the dinghy will quietly rise up the wave and slide down the other side, and you will maintain good steering control whether under oars, motor, sail or even sculling.

Similar problems can arise when landing on a beach if there is any surf rolling onto it. The best way to handle this is to turn the dinghy round and go in stern first. The waves will tend to carry you in and you can use the oars

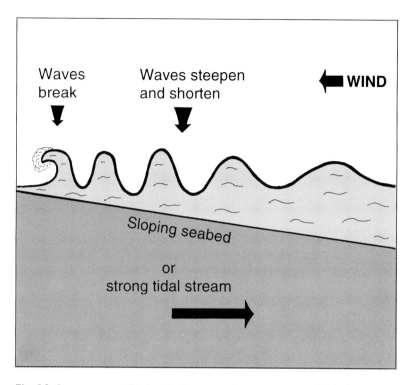

Fig 18 As waves roll into shallow water or run against strong tidal streams (see next section), so the bottom part of the waves are slowed by the friction. This causes the waves to bunch-up as you can see, and eventually break when they become too high to be stable.

to control your speed and direction. Do not do this with an outboard motor or the propeller will land on the beach first and cause the dinghy to capsize. The moment you ground on the beach, hop out and pull the dinghy quickly up and clear of the waves. If you go in bow first the waves will more easily pick up the blunt stern than the sharp bow and they can then swing the boat broadside and sweep it ashore upside down, or at least full of water. If you do beach bow first try to nip in quickly between waves, then hop out and drag the dinghy rapidly up the beach before it can be swamped. See Figure 18.

Wind and waves also affect the yacht when at sea, as you can imagine, and it is important that the skipper be able to predict the weather before setting sail. Special weather forecasts for sailors are broadcast regularly on the radio, and full details will be found in Appendix 1. One of your jobs when crewing on the yacht may very well be to note down the 'Shipping Forecast', as it is called. Recording this accurately is extremely important and a skipper will be delighted to have a crew who can be relied upon to do so. Study the appendix carefully and practise listening to and writing it down at home. It is often read out very rapidly, especially if it is complex, as only five minutes are allowed for it, so practice is essential to be able to record it accurately and completely.

Tides and Currents

Tides are movements of water in harbours and the sea caused by the gravitational pull of the Sun and the Moon. The level of the water is very high twice a day (*HIGH TIDE*) and very low twice a day (*LOW TIDE*). Thus if High Tide is at six o'clock in the morning, the level of the water will gradually drop until Low Water (or Low Tide) at about twelve noon, then rise steadily again to the next High Water at six in the evening; and so on. See Figure 19.

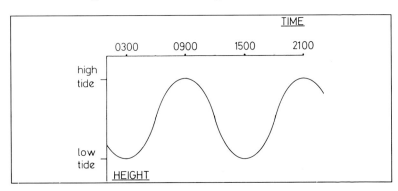

Fig 19 This tide is low at three o'clock and high at nine, morning and night

When the tide is rising and falling it also moves along, and will carry with it anything that floats on it, such as you and your dinghy. When it rises it flows into harbours and estuaries, and when it falls it flows out. The first is called the *FLOOD TIDE* and the second the *EBB TIDE*.

The tide is not the same every day, however, the continual movement of Sun, Moon and Earth causing a progressive change in both its time and its height. The rotation of the moon around the earth causes the tide to be nearly an hour later each day, while the rotation of both around the sun causes the *RANGE* of the tide (difference between heights of high and low water) to vary throughout each month. See Figure 20. Even the twice-daily change from high to low water is not regular, as you can see in Figure 21. Currents are also flowing water, but as they are not caused by Sun and Moon the water does not alter in height or speed. The flow of a river is a typical example.

If you are out in a dinghy when the tide is running, especially halfway between High and Low Tide when it runs the fastest, you must take great care to steer so as to make allowance for where the tide will be setting you, or you could find yourself drifting into danger or out to sea. In Figure 22 you can see how this is done. In certain places at *SPRING TIDES* (see Figure 20) the TIDAL STREAM can run far faster than it is possible to row, so it is not safe to go out in a dinghy at such times until you are extremely experienced.

You should also watch for the effect of a tidal stream running towards a strong wind. The stream pushing against the wind will exaggerate the waves, causing them to be steeper and higher than normal. See Figure 18. When the tide turns, the effect of the stream running away from the wind will be to reduce the height and steepness of the waves. Watch this a few times and you will soon get the feel of how seriously this could affect the safe handling of a dinghy. The difference between the waves when the wind is against the tide and when it is with the tide can be very marked; so much so in strong winds that you should consider delaying dinghy trips until wind is with the tide.

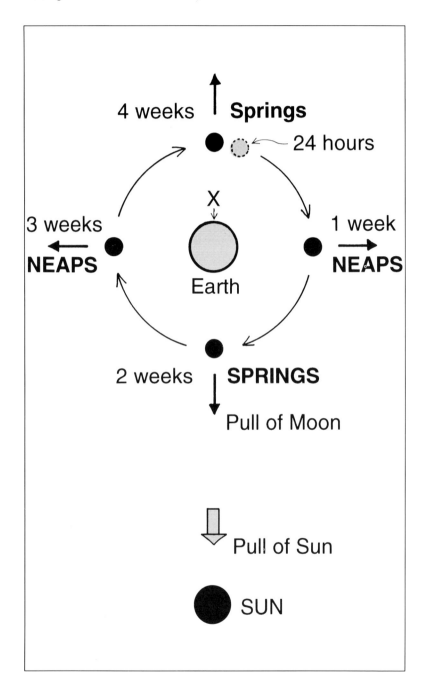

Photo 16 This yacht is moored in a river and is free to swing round its mooring. As the flood tide runs into the river so the boat will swing to lie with its stern pointing upriver, the tide sweeping the boat away from its mooring chain as you can see.

⇐

Fig 20 When the sun and moon are in line their pulls are added together, creating very high High Tides and very low Low tides, known as Spring Tides. These occur every two weeks, a couple of days after Full Moon and New Moon. On the weeks in between the pulls are at right angles, producing relatively low High Tides and high Low tides, known as Neap Tides. The range of Spring Tides is much greater than for Neap Tides, and it changes gradually during the week between Springs and Neaps. During the 24 hours it takes the point 'X' on the Earth to go right round and back to its starting point, the moon moves as shown. This means that the Earth must rotate a little further in order for point 'X' to get under the moon and thus produce the next day's High Tide, and this is why the tide is about fifty minutes later each day.

51

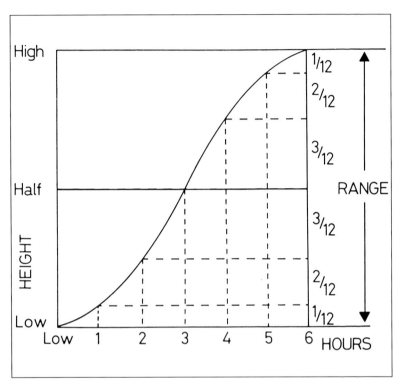

Fig 21 This diagram shows a simple tidal calculation called the 'Twelfths Rule', that can be used to work out the approximate height of the tide at any time. If you look carefully you will see that the tide rises (or falls) one twelfth of its range during the first hour, two twelfths during the second hour and three twelfths during the third . In the fourth hour it changes by another three twelfths, then two twelfths in the fifth and one twelfth in the last hour.

Fig 22 ⇒
You can see from fig 22a the course that the boat makes good while steering across a tidal stream that pushes it sideways. Any two objects ashore, such as these two beacons, can help you judge how the stream is setting you. To maintain this course made good you need only steer so as to keep the two beacons in line; if the distant beacon(A) opens out to the left of (B) then you are to the

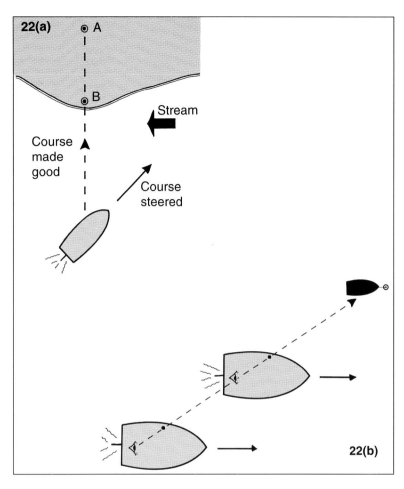

left of this track, if it opens to the right you are to the right. By watching marks such as these you will be able to steer a course that will make the boat make good the direction you want to go in. Much the same can be done by lining up a mark ashore (or moored yacht, perhaps) with an object on your own boat.(Fig 22b) As long as your eye stays in the same position in your boat, this will indicate your course made good. If the mark on the boat stays in line with the distant mark, the boat will move towards that mark, as you can see. If the mark on the boat moves to the right of the other, so will the boat.

Photo 17 This shows how the effect of tide running past a buoy indicates its speed and direction. You can see clearly that the tide is running from the right of the picture to the left - and quite strongly.

It is quite easy to check which way the tide is running in a harbour, simply by looking at boats which are *MOORED* just by the bow. Because they are free to swing round, the tide will cause them to lie away from the *BUOY* in the direction it is running. See Photo 16. In the absence of moored boats you should be able to see the water actually flowing past buoys, beacons, jetties etc. See Photo 17. It is best to check this away from the shore as the tidal flow is often disturbed by the presence of jetties and the shore itself. Generally the stream is strongest in the middle of the deep channel, growing weaker in the shallows close to shore. Sometimes very close inshore there is even a *BACK EDDY* that runs in the opposite direction to the main stream. See Figure 23. You should, therefore, keep close inshore when rowing against the tide and out in the main channel when going with it.

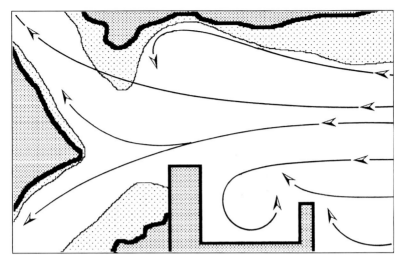

Fig 23 The arrows show how the tide is likely to be affected by obstacles such as bends, jetties etc.

Details of tide times and heights are contained in *TIDE TABLES*, which should be on board the yacht. Times and heights are usually listed for each day of the year so it is a simple matter to see when the tide is high that day, and whether it is near *SPRINGS* or *NEAPS*. You should be able to calculate the difference in height between high and low water that day, and from this make an assessment of how far the tide will rise or fall while the dinghy is moored up. See Figure 21. In the summer you must check whether the times given are Summer Time or GMT (Greenwich Mean Time, which is kept in the winter), or you could be an hour out in your calculations.

You must also check the harbour for which the tides are predicted, as it may not be the one you are in. If not, there should be a list at the back of the tide table of the corrections you must apply to the listed times to get the correct times for the tides in other nearby harbours. The reason for this is that the tide runs along the coast much like a wave, producing High Water at gradually later times as it reaches the harbours one after the other. High Water will also be later at the top of a long estuary or river than it is at the mouth, sometimes by as much as some hours. Times and heights may also be affected by weather conditions and wind, and the skipper should brief you on possible effects in your area. Some thought needs to be given to the interpretation of tide tables.

Chapter 5
Anchoring and Mooring

The anchor and warp can be used to secure the dinghy to the seabed to stop it drifting about. You can do this to have lunch or whatever, or if you get in trouble. If you lose an oar, for example, and find the tide or wind carrying you into danger you should anchor immediately, then attend to sorting out the problem. If you cannot rescue yourself , you can use the signals given in Chapter 1 to call for help. See also Appendix 3.

Anchoring the Dinghy

Basically simple anchoring technique can be seen in Figure 24. Drop the anchor over the side and pay the warp out steadily as the dinghy drifts on the tide or wind. This will ensure that the warp does not pile up in a heap on top of the anchor, where it might catch on a *FLUKE* (the pointed parts on an anchor that dig into the bottom) and pull the anchor out when the weight comes on it. In theory you should let out approximately five times as much warp as the depth of water (see Chapter 9), but in practice with a dinghy we generally simply let out all the warp. If you want to gauge the depth, you can lower the anchor to the bottom and estimate the length of warp needed. See Photo 18. The end should be permanently secured to the *RINGBOLT* in the *STEM* or the for'ard thwart, or round the mast if you have one, using a suitable knot (described further on). Put the warp into the for'ard *FAIRLEAD* (U-shaped fitting on the gunwale) so as to make sure it leads from the bow and thus holds the dinghy head into the wind or tide.

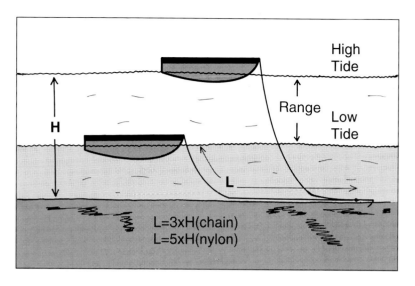

Fig 24 You can see how the anchor warp should be laid out in a line so that it pulls evenly on the anchor. It should not be dropped in a heap on top of the anchor as it may then tangle and pull the anchor out when the weight comes on. Note that a chain warp must be three times the depth and a nylon one five times. This refers to the maximum depth the water will be while you remain at anchor, so you must check the tide times and calculate from the 'Twelfths Rule' how much deeper or shallower it will become. See Fig 21.

Photo 18 This dinghy is anchored by the stern and secured to the shore with a very long bow painter. This system holds the boat clear of the shore. See Fig 25.

To make sure the warp runs out smoothly when you anchor, which can be important if you have to anchor in a hurry, you should always keep it neatly coiled in the bottom of the dinghy. Start the coil with the end secured to the dinghy and finish with the anchor on top. There should be a couple of metres of chain between the anchor and the warp, which helps protect the warp from chafing on the bottom and also adds a bit of weight to stop the anchor lifting out. To check that the anchor is holding firmly in the bottom you should feel the warp just above the water. If the anchor is 'DRAGGING' (sliding along the bottom), you will be able to feel the vibrations in the warp.

If you are not anchoring because of an emergency then you should try to do so somewhere sheltered from the wind and waves, and also out of the main run of the tide, as you will be a great deal more comfortable if you can. Do not, however, anchor where there appears to be an obstruction, such as a wreck or rocks, that you might ground on if the tide is falling, and keep well clear of the main channel where yachts and ships will be sailing. Unless you specifically want to dry out you must check the depth and calculate how far the tide will fall while you are anchored, so as to ensure that you remain afloat. If the tide is flooding, then make sure you let out enough warp to allow for the extra depth later when the tide has risen higher. Before ever going out in the dinghy, then, you should consult the tide tables so that you can always work out in your head what the tide is doing at any time - even if only approximately.

If you pull the dinghy up on a beach you should carry the anchor away up the beach and bury it in the sand to ensure that the dinghy cannot float away. On a SLIPWAY the boat should be tied to a nearby post with a long bow painter, even if it is well clear of the water. The tide may cause the water to come higher up the beach later, so try to bury the anchor or secure the painter far enough up beach or slipway that they will be always above the water. This is particularly useful when the dinghy is too heavy to drag up the beach. See Photo 19.

Photo 19 The crew taking the dinghy's anchor up the beach and burying it securely in the sand. The tide will have to rise a long way to cover this anchor, as you can see from the slope of the beach.

Mooring the Dinghy

When going alongside a yacht or jetty you should head into the tidal stream so that it will help you stop on arrival. You should then secure the painter as quickly as possible, before the tidal stream sweeps you away. Going alongside 'down-tide' (with the tide pushing you along) is much more difficult, and can be dangerous at times as it is so much harder to control the dinghy. If the wind is very strong you may have to head into it instead. If in doubt, aim to berth pointing towards the bow of the yacht if she is on a swinging mooring. Practice is needed to determine how soon to stop the outboard or pull the oars inboard and still continue to coast right alongside the yacht, without crashing into her. Swing the dinghy at the last minute so that it runs alongside the yacht as it slows to a stop.

The other effect that tides will have on you in the dinghy occurs when you moor up to a jetty or land on a beach. If you do this at or near Low Water you must secure the painter to a place that will be above the High Water level or you will not be able to reach it to untie the dinghy when you come back! It might even hold the dinghy down and sink it as the tide rises. You must also take care not to moor the dinghy in a place where it could float up with the tide and get caught under a jetty or similar. Conversely, if the tide is falling when you moor up you must leave a long enough painter to allow the dinghy to go down with the tide to Low Water. If necessary untie the anchor and use the anchor warp. You should be able to tell where the tide will rise to by a line of seaweed and general rubbish along a beach or green slime on a slipway or the side of a jetty. Two distinct lines should show the Spring and Neap heights.

If the dinghy is to be moored to a pontoon or jetty it should be tied up with the bow painter only so that room is left for other dinghies to moor. Try not to lead the painter across rough or sharp edges on a quay in case the dinghy roams around and drags the painter across it; if left like this long enough the painter could part and lose you the dinghy. A neat system, known as the *HYPOTENUSE MOORING*, can be seen in Figure 25. The anchor is thrown out astern and its warp secured at such a length that the dinghy will just reach the shore. The bow painter is then taken some way to one side along the shore and secured so that the dinghy lies at an angle, which both keeps it clear of the landing steps and prevents it banging into the jetty.

Out at the yacht the dinghy can either be secured astern so that it drifts clear, or tied alongside with fenders in between, using both the bow painter (to hold the bow in) and the stern painter (to hold the stern in). Leave a little slack in these painters so that the dinghy can rock about in waves without snatching too hard. See Photo 20. If there is a boarding ladder over the side of the yacht then keep the dinghy well clear of it to prevent damage or tangles if the waves are rough and cause the dinghy to bounce about.

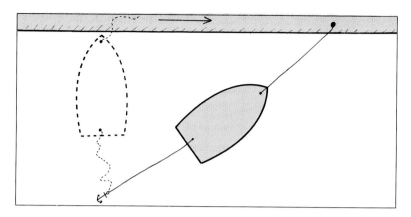

Fig 25 The hypotenuse moor.

Photo 20 A collection of dinghies secured in various ways. In the foreground is a traditional wooden sailing dinghy anchored. Beyond it is a lightweight plywood sailing dinghy moored alongside a yacht. Beyond and to the right is another dinghy tied to a yacht's mooring while the yacht is away sailing on a short trip. In the background is a larger dinghy moored astern of the big sailing barge that is anchored in the middle of the river. Which way is the tide running?

If the dinghy is secured astern of the yacht you may find that in certain conditions the wind will blow it forward to bang against the yacht. To reduce the likelihood of this you can moor the dinghy on a very long painter so that it floats well clear. Better is to hang a bucket over the stern of the dinghy (one with a strong handle) so that the tidal stream - into which the yacht will almost certainly be pointing - will drag on it and pull the dinghy clear of the yacht.

When transporting stores to and from the yacht care must be taken to prevent them falling over the side. Keep valuable items tied to the dinghy while under way, and secure bulky stores to the yacht before passing them into or out of the dinghy, so they can be retrieved if they are dropped. Take great care to keep the dinghy steady and properly trimmed when passing heavy things across.

Securing the Dinghy

It is important when securing anchor warps and painters, buckets and what-have-you, that a suitable knot be used. The knot must hold firmly and reliably even when the rope is thrashing about, but also must not pull tight and jam so that you cannot undo it. It must also be possible to untie the knot when it is under strain. The only practical knot with all these qualities is the *ROUND TURN AND TWO HALF HITCHES*, and it can safely be used for virtually every task you will come across on a boat. See Figure 26. The knot should be tied tightly, especially if it is being put into a thin metal ring, in order to prevent movement which could chafe through the rope.

Another knot you may find useful at times for mooring the dinghy is the *BOWLINE*. This knot produces a loop of whatever size you need and you can see how it is tied in Figure 27. The Bowline should not take the place of a Round Turn and Two Half Hitches unless a loop is specifically needed - to slide up and down a post perhaps - as the loose bight can rub and chafe. Also it cannot be tied or untied when the rope is under strain, as a dinghy painter may well be in strong winds, or if the dinghy becomes wedged behind another. If another bowline is

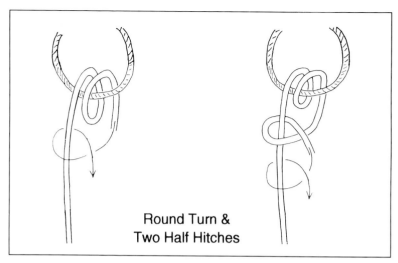

Round Turn &
Two Half Hitches

Fig 26 Pass the working end twice round the ring, as in the left hand picture, to make a complete 'round turn'. Then 'half-hitch' the working end round the standing part as shown by the thin arrow and draw the knot up tight. Finish with a second half-hitch as shown by the thin arrow in the right hand picture.

already on the post you wish to use you must 'dip' yours through it, so that either one can be removed without disturbing the other. See Photo 21.

Unless the loop can be lifted off a post without having to be untied this knot should not be used. Many books and even more people will recommend it to you as being 'the most useful knot of all'. Do not believe any of them; it is not. It has certain, very limited functions for which it is admirable, but it does not fulfil a single one of the requirements of a general purpose knot that are listed above. The Round Turn & Two Half Hitches is infinitely superior.

You should also know how to tie the Sheet Bend as this is probably the best general purpose knot for joining two ropes together, even if they are of different thicknesses. Get into the habit of always passing the end round twice, making a Double Sheet Bend, as this is very much more secure. See Figure 28.

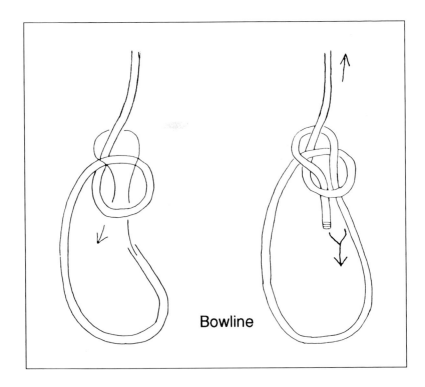

Bowline

Fig 27 Form a small eye in the standing part about twice as far from the working end as the diameter of the bowline required. The standing part should emerge from the back of this eye, so that the finished knot holds it tightly against the eye: see left drawing. Then feed the working end round as shown by the thin arrow and tighten the knot (right-hand picture) by hauling up on the standing part and down on the working end and the right side of the loop together (see arrows). Make sure all parts are snugly together and that a few centimetres of the working end protrude so that the knot cannot shake loose.

Photo 21 The second bowline to be placed on this mooring bollard is passed up through the eye of the first, then slipped over the post. As you can see, either can be removed easily without disturbing the other (which may be under strain and incapable of being slackened).

All these anchor warps and painters will get very untidy if you are not careful, and if that happens you can be sure they will tangle and jam just when you need them in a hurry. Any rope on a boat, small or large, should be coiled up neatly when not in use and stowed carefully in the bottom of the boat. Then, if you need to anchor in an emergency for example, you can be sure the warp will run clear. See Photo 22.

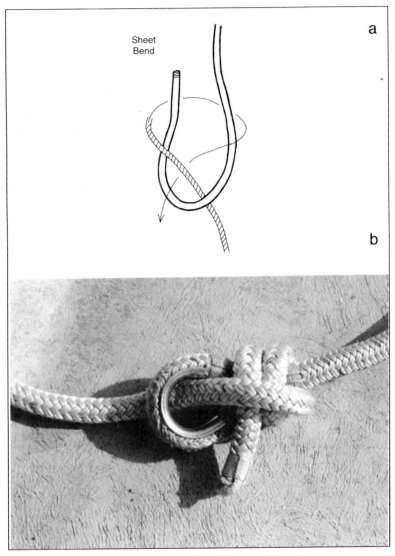

Sheet
Bend

a

b

Fig 28 Make an open loop in the thicker of the two ropes and
pass the other round as shown by the long arrow. Both ends
must protrude on the same side of the knot and be long
enough to ensure it cannot shake loose. If a double sheet
bend is required (see photo) then pass the end round again
before working the knot tight.

Photo 22 This rope is being coiled clockwise, starting with the end that is attached to the cleat. A painter in the dinghy can be simply dropped neatly in the bottom of the boat after coiling. For the coiling and stowing of other ropes on board the yacht see Chapter 6.

Foredeck work, such as slipping the mooring when sailing off, will constitute much of your work when crewing the yacht. Smooth efficient work up here will greatly simplify the skipper's job.

Part Two - Crewing the Yacht

Chapter 6
General Seamanship

Although in principle crewing the yacht is not really different from pottering about in the dinghy, in practice it is; mainly because everything is very much bigger, more complex and under a lot more strain. There is greater potential for danger and damage than there is when handling a dinghy, so sailing a yacht requires more organisation, in order that all the crew work together as a team. The person who does this organising is the skipper. He is in charge of the yacht and her crew, and is responsible for their safety. In the old days when sailing ships were at sea for months on end, with no radio on board with which to call for help or advice, sea captains were known as 'Masters under God', because they were answerable to no-one but Him. Everything that happened on the vessel was the captain's sole responsibility, and on a small yacht even today much the same applies.

It is essential that the skipper's orders are obeyed without question, and that he can rely on his crew absolutely, if the boat is to be handled safely. Reliability is probably the most important quality required of a crew, but it is closely followed by willingness, and cheerfulness in the face of difficulty. Pottering about in the dinghy will certainly help to draw out these characteristics, as well as instilling in you the basics of good seamanship, but the more detailed and specific aspects of the subject will need to be studied, and the second part of this book should help you with that.

My dictionary defines seamanship as: skill of a good sailor; which means a great deal more than simply being able to sail well. Virtually all facets of life onboard a boat are quite different from their counterparts ashore, even ordinary domestic tasks like washing up requiring a methodical and determined approach when conditions are difficult. Out at sea in rough weather when you can become exhausted simply from standing still on a crashing, rolling, pitching boat, or when you are so utterly tired from lack of sleep that you cannot imagine how you will stay awake ten more minutes to the end of your watch, the simplest, most trivial task can seem quite impossible. The good seaman is a sailor who handles these difficulties and differences as though they were a perfectly normal, everyday part of his life.

A counsel of perfection this undoubtedly is, but it is what you as a crew should strive for. If you can develop the right attitude, the necessary cheerfulness and determination to never give up, then you are well on the way to being able to make real use of the skills that we shall now discuss.

Handling Ropes

A tremendous amount of a sailor's time is spent handling ropes of one sort or another and a sign of a good seaman is the efficient way he does so. As ropes under tension can be extremely dangerous because of the tremendous forces unleashed if they slip or break, it is vital that you learn to handle them confidently and safely, keeping them well under control at all times. To do this you need to understand what is happening when the strain comes on a rope. There are various different types of rope used on board a boat, and they all have different properties and therefore uses and handling techniques. Natural fibres such as hemp and manila are rarely used nowadays, being generally difficult to handle and liable to rot, and most ropes will be of man-made materials.

Terylene™ (or Dacron™) is perhaps the most commonly encountered, being very strong, stretch-free and generally comfortable to grip and handle. It is probably the most versatile of all ropes, being excellent for *HALYARDS* (ropes used to hoist the sails), sheets, mooring warps, lashings and so on, but is expensive. Nylon is similar but stretches considerably when under strain - perhaps to one and a half times its original length. Because of this its use is limited to applications where stretch is useful, such as for anchor and mooring warps. The stretching capability is then utilised to absorb sudden shocks on the warp, caused by the boat pitching and straining when in waves. Polypropylene is a cheaper substitute for Dacron/Terylene, but it is not so strong and it is rougher and less comfortable to handle. It also chafes more readily. It does, however, float, which the others do not, so has its special uses.

This business of chafe can be a serious problem on a boat, and great care needs to be taken to prevent it occurring, particularly in seriously important ropes like mooring and anchor warps and so on. Check carefully that ropes do not press or rub against sharp or rough objects, either on the boat or on the shore. If there seems a risk of chafe the rope should be adjusted to lead differently or plastic hose or similar fitted on the rope to protect it. See Chapter 9.

Most rope is laid up in twisted strands, although sheets (the lines controlling the angle of the sails) are often braided (see Photo 23), as this construction is soft to handle. Anchor warps are best plaited as this type is enormously strong and chafe-resistant, and has no tendency to twist. See Photo 24. When not in use ropes must be coiled and stowed away so that they are always ready for instant use without tangling. See Figure 29. The spare ends of halyards should be coiled and hung on the *CLEATS* to which the rope has been secured. In Figure 30 you can see how to turn up a rope round a cleat and Photo 25 shows how to then stow the spare end of coiled halyard.

In Chapter 5 we coiled ropes in the dinghy loosely clockwise and simply dropped them on the bottom boards so that they would run without tangling. Most of the time this simple clockwise rule serves well enough, but the question of which way to coil a rope is actually quite complex, and sometimes important. In most laid rope the *STRANDS* are twisted from left to right and the rope should be coiled clockwise to prevent kinking. If strands are laid right to left (very rare nowadays), the rope should be coiled anti-clockwise. Braided and plaited ropes have no tendency to twist so should be coiled loosely in figures of eight so that no particular overall opposite twist is put in.

Photo 23 Note the difference in appearance between the laid rope on the left and the braided rope in the hand.

Photo 24 This is plaited rope, ideal for mooring lines and anchor warps as it is immensely strong and not prone to twisting.

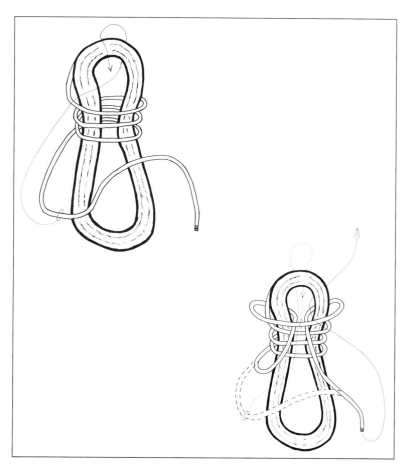

Fig 29 After the coil is made take a long end and turn it tightly around the body of the coil three or four times as shown on the left. If the coil is to be hung or just stowed temporarily, you can finish very simply by poking the end right through the eye at the top of the coil as shown by the thin arrow. Pull tight and use the end to tie the coil to a rail or wherever. For a more secure stow pass just the bight through (leaving the end on this side), then pull the bight forward over the top of the coil, as on the right, and pull the end tight. Follow the dotted arrow line. Finish by passing the end through the eye as shown by the thin continuous arrow.

Fig 30 Note how the rope leads in from the side of the cleat; this keeps it clear of the turns. After the final turn the end can be tucked under the last turn as shown, to make the turns more secure. This is especially important with springy rope.

Photo 25 Stowing the *FALL* of the halyard by simply wedging the coil between the taut halyard and the mast, just above the cleat (behind the man's arm).

If a rope does become kinked, the kinks should be gently untwisted starting from the fixed end of the rope (if it is attached to the boat) so that they can be eventually wound out of the loose end. A good way of doing this is to throw all the rope over the side and haul it in slowly, recoiling properly as you go, beginning with the fixed end so that the loose end is on top. Take great care not to get it caught round the yacht's propeller or keel.

It is important to realise that many of the ropes and lines that you handle onboard the boat are likely to be under considerable tension, and if handled incorrectly can cause damage to both the boat and the crew, as well as making a mess of the current manoeuvre. Nylon, in particular, if stretched under great strain, will lash out like a whip if it breaks or is released suddenly. Apart from the risk of breakage (extremely remote with the enormous strengths of modern materials), there are two basic potential problems: the possibility of a rope you are holding under strain becoming too hard to hold; and the virtual certainty that given half a chance any rope that is slack will wrap itself firmly around every obstruction it can find.

When handling a rope that can come under strain - mooring warps and sheets mainly - always be ready to take a rapid extra turn round the winch, cleat or bollard if you feel increased strain coming on - before the strain gets too much for you to hold. These turns must be wound on carefully so that control is always maintained. If a rope once starts to run away with you it will be extremely difficult to regain control, not to mention dangerous. If a rope needs making fast quickly - a stern rope to stop the boat for example - then do not waste time trying to tie a knot but just get a few turns rapidly round a handy bollard or cleat and lean back hard against it to take the weight. Keep your fingers well clear of where the warp is turned up - cleat, winch or whatever - in case a sudden jerk makes the rope slip and pull your fingers in.

Putting an extra turn on a winch or bollard under load can be fraught with risk, as it is difficult to keep strain

Photo 26 To put an extra turn on a winch take the weight with your left hand and lay another turn loosely round the winch (above the existing turns) with the right, as shown here. The loop should be much smaller than this, with the left hand just far enough from the winch to prevent it being pulled in suddenly and jammed between rope and winch. The rope is then allowed to slip off the fingers of the left hand, at the same time rapidly hauling tight with the right. With practice, this will enable you to put turns on very quickly and perfectly safely in the toughest of conditions. Practise it without strain on the sheet until you become confident.

on the working end while at the same time turning it round the barrel. If there is enough space, you can simply wind it round while leaning back hard to hold the weight - keeping your fingers well clear of the turns in case the rope slips a little. If you have no room to safely do this without leaning over and risking losing balance, then a turn can be put on as shown in Photo 26. Done properly, this is an extremely useful method of getting extra turns on very quickly and safely; practise it with light loads that pose no danger.

It is equally important to control and keep the strain on a rope while easing it out, and this can be difficult to do smoothly if there are too many turns on a winch or round a bollard. Turns can be assisted to render round a winch by pushing them round with the palm of the hand, keeping the fingers well clear, but there is no safe and simple way to remove a turn - it must be carefully unwound while keeping control of the strain. You may see very experienced seamen quickly flick the top turn off then heave back to hold the strain, but do not try it until you are equally experienced!

It is important that you know the correct way to turn up a rope on a cleat (see Figure 30), and also how to tighten it by *SWEATING*. Not only must the rope hold securely on the cleat, but it must also be easy to release, and gradually ease off, if it is under a heavy load. To tighten a rope that leads onto a cleat you should simply hook the bight of the rope under the bottom horn of the cleat and hold the rope firmly so that it bears against the cleat hard enough to prevent it slipping. Then yank out sideways on the rope a few feet above the cleat while holding it firmly against the cleat; this will generate some slack in the rope above the cleat. Let go of this slack suddenly and at the same time heave on the end to pull the slack quickly round the cleat before the strain comes back on. It is a bit of a knack, but will quickly come with practice. See Photo 27.

Finally, you will find from time to time that the ends of ropes become loose and frayed. Frayed ropes can be extremely dangerous on a boat as the ends can so easily get caught up and jammed. There are many ways of preventing this but the quickest and simplest will do for the moment, although it is only a temporary solution. Take a roll of insulating tape, tuck the end under a strand then wrap the tape half a dozen times tightly round the whole diameter of the rope, working with the lay of the strands. Cut off the frayed end of the rope close to the tape. If possible, seal the end of the rope by holding it over a flame so that the strands melt and fuse.

Photo 27 To tighten a halyard by sweating, hook the halyard under the cleat and hold very firmly while yanking out sideways on the halyard above the cleat. Let go with the left hand and heave the slack rapidly round the cleat with the right.

Knots and Lashings

The tying of knots properly is a vital part of sailing: hardly a moment goes by without one being needed for some reason or another. When used on boats it is important to appreciate that considerable and complex strains can often come on knots, so the correct one must always be used for a particular purpose. Although we mentioned in Chapter 5 the importance of using the correct knot for a particular task, you may have found that it does not seem to be too critical in practice. On a big yacht, however, the strains and stresses are such that a wrong knot could cause a disaster. Let us look at the common uses for knots on board the yacht and see which knots will do those jobs best.

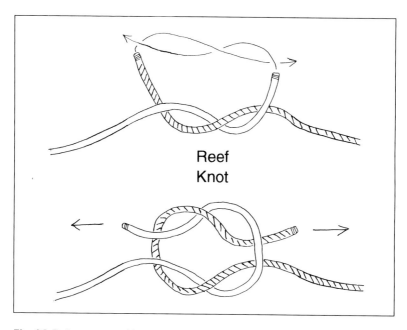

Reef
Knot

Fig 31 Take an end in each hand and pass the left one over the right one, down the back of it and up the front, as in the top picture. Then bring the ends together again and pass the new right one (originally the left) over the left, down its back and up its front so the knot looks like the bottom picture. Draw the ends tight.

Tying the boat up to a buoy or a jetty is perhaps the first task to spring to mind, and the best knot will depend on the type of fastening point available. Generally, the Round Turn & Two Half Hitches should be used, unless large bollards are available in which case a Bowline big enough to drop over the bollard will do well. See Figures 26 and 27 in Chapter 5.

The Reef Knot is a well-known knot, much in use ashore for joining two ropes together. At sea, however, it should never be used for this purpose as it can capsize under strain and fall apart. It is used for tying reef points in a sail as explained in Chapter 7, and for no other purpose except wrapping up birthday presents. See Figure 31.

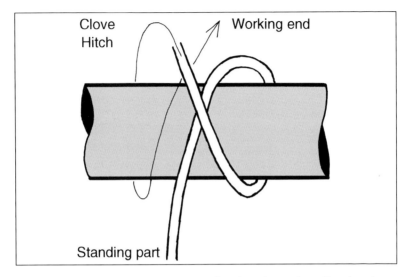

Fig 32 Take the end over, down the back and up the front, then cross it over the first part, down the back and up the front again underneath itself. Pull the ends tight. This is a useful knot for tying a fender to a guard-rail as it can be quickly adjusted by loosening the knot and pushing the end through to raise or lower the fender.

The Clove Hitch is a useful knot when used in its proper role of a 'crossing knot', but it is often vaunted as just the thing for tying up dinghies and various odd jobs when the Round Turn & Two Half Hitches is very little more trouble and infinitely more secure. Probably the most useful purpose of a Clove Hitch is for lashing the tiller, when a line can be secured on one side of the cockpit, clove-hitched to the tiller, then the end hauled taut and secured on the other side of the cockpit. It should never be used for mooring a boat, however small. See Figure 32.

When a boat is rolling and leaping about in waves it is most important that all gear should be securely lashed to prevent it flying around. Large objects on deck, such as dinghies, can be lashed down very tightly using Waggoner's Hitches (often used for securing loads on lorries). See Figure 33. Another useful way of tightening a lashing is by *FRAPPING* (hauling it sideways with

80

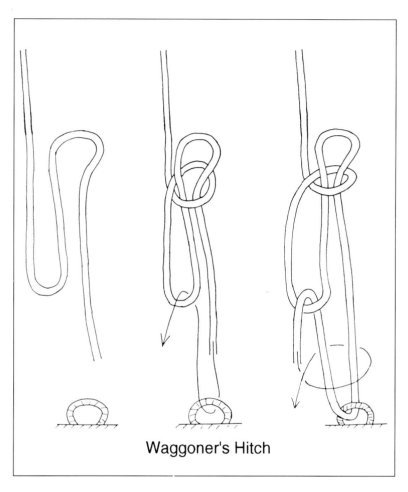

Waggoner's Hitch

Fig 33 This is a most useful knot for tightening a deck lashing holding a dinghy perhaps. It can be made more secure by lashing the top small eye sideways to the standing part so that it cannot slip through. Form the eye as on the left , leaving enough room below the bottom loop to pull it down when it tightens. Form a small eye in the standing part and noose it tightly round the top loop to make an eye as in the middle, then pass the end through an eyebolt etc on deck, up through the bottom loop and pull it down as tightly as possible.Finish with a half-hitch as shown on the left, and add a second for security.

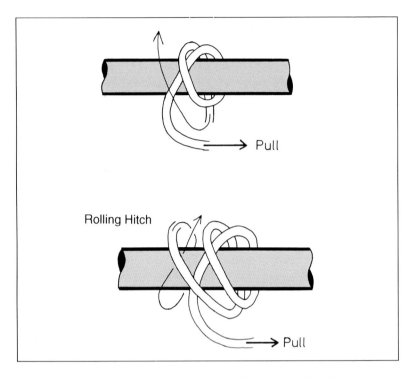

Fig 34 Start as for a clove hitch (Fig 32), laying the first cross away from the direction of pull. If the pull was to the left along this spar, the first turn would come up the front on the left of the end taking the strain, and the subsequent knot made towards the right instead of the left as here. Tuck a second turn slightly wedged between the first turn and the end taking the strain (standing part) and finish with a tuck on the opposite side of the standing part, as in the bottom picture. The knot must be made tightly as you go along, not loosely then pulled tight at the finish.

another line).   Finally, let us look at a much under-rated knot which enables you to haul sideways along a rope, spar or whatever - the Rolling Hitch. See Figure 34. There are countless uses for this knot, but perhaps the most important is for taking the strain on a jib sheet that has jammed with *RIDING TURNS* on the winch. See Chapter 7.

Safety Precautions

The business of safety at sea is made much of these days, so much so that it can obscure the basic precautions that a good sailor thinks of as no more than elementary seamanship. A sailor needs to be self-sufficient when he is at sea and it is important that he think positively about how to deal with trouble if it crops up. By and large this is the skipper's concern, naturally, but there are certain ways in which the crew can help - mainly by simply being aware of the troubles that could befall the yacht and of how the skipper is likely to deal with them. There are four main potential emergencies that could confront a skipper: fire; sinking; man overboard; and medical distress. All, if handled properly by a competent skipper, can usually be prevented from becoming genuine emergencies; especially if an efficient, well-briefed crew is on hand to assist.

A good skipper will have carefully-thought-out routines for each of these situations, posted clearly somewhere for the crew to read. He will also equip the yacht with the correct gear for dealing with them, and brief you all on using it. See Figure 35.

```
                    Emergency Routine

                          Fire
1 Shout FIRE !FIRE! FIRE!
2 Alert all hands
3 Attack fire with suitable extinguisher
4 Turn boat so smoke blows clear
5 Switch on VHF - emergency Channel 16

      Read Fire Extinguisher Instructions
```

Fig 35 An example of an emergency routine in case of fire. The essence of these routines is to provide very simple basic instructions to get a potentially dangerous situation under control quickly, enabling you to make time to think about the particular problems that need dealing with.

Fire: Fire extinguishers will be installed in various places throughout the boat, situated close to likely fires but also where the fires cannot prevent you from reaching them. See Figure 36. Note carefully where they are and read the instructions on how to operate them, and remember both; it may be dark when you need one.

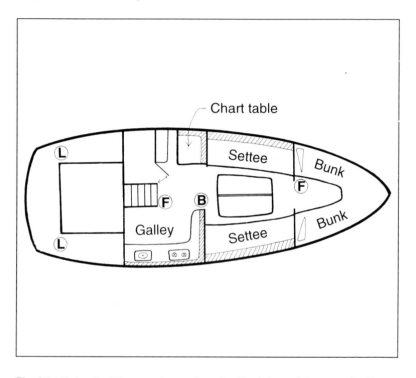

Fig 36 Lifebelts (L) are placed on both sides of the cockpit where they can be reached easily. A fire blanket (B) is situated close to the cooker so that it can be placed quickly over a burning pan etc to put the flames out. Fire extinguishers (F) should be put where they can be reached from on deck if possible (main hatch and forward hatch) in case of a raging inferno below, and also close to possible fire sources, but not positioned so that you would need to reach over the potential hazard. The one in the forecabin can be reached from forward without having to pass a fire in the galley. The after one is close to the engine.

Sinking: Note the positions of *LIFEBUOYS*, life-jackets and *LIFERAFT* and find out how to operate them; the skipper should brief you on all this, and also on how to fit and when to wear a *SAFETY HARNESS*. This latter is a strong harness that fits round your body and can be clipped to the boat to prevent you being washed or knocked overboard in rough weather. See Photos 28a & b.

Photo 28a This 'horseshoe' lifebelt is at the stern, easily reached from the cockpit by the helmsman should anyone fall over the side. It simply lifts out of its support, along with the light that automatically begins flashing when it lands in the water. There are many other lifebelt systems, probably the best having a belt on each side of the cockpit rather than just one on the stern. Two are then available, and both can be reached more easily than the one shown here. See Figure 36.

Photo 28b The webbing harness fastens securely round the shoulders and upper chest. The line has a hefty spring clip that can be hooked over special wires called jackstays that run the length of the boat.

Man Overboard: The lifebuoys should be close to the
helmsman so that he can throw them over after a man. At least one should have a light attached, so make sure this is the one you throw in the dark. Many boats have *DAN-BUOYS* with tall poles on them which can be seen for a very long way, and they often have a quick release system. Check how it works. If someone does go overboard the instant release of a lifebuoy could well save his life, so make certain you know how to do this.

Somewhere near the hatchway should be some *FLARES* for use in various types of serious situations. White ones are used to alert other vessels to your presence (usually big ships coming too close that may not have seen you - see Chapter 8), and red ones can be used to summon assistance if the yacht is in very serious trouble. There are two types of these, small ones that can be held in the hand and large ones that fire rockets with red stars very high in the sky. The skipper will doubtless brief you, but as with the fire extinguishers, do remember that if you ever need them conditions are likely to be too bad, or too urgent, for you to sit down and read the instructions! Read them regularly and remember them. Note which way round the flares should be held, and remember where they are stowed.

Two things that could cause fire onboard are gas and engine fuel (especially if petrol). The skipper should brief you very carefully on how to deal with these things as safely as possible. It is unlikely that you will be concerned with fuel, but operating a gas cooker will almost certainly be within your domain. See Chapter 10 for information on how to deal with this most important aspect of life onboard a yacht. Note carefully the *BILGE PUMPS* - powerful pumps for removing water from the bottom of the boat (bilge) if she leaks. It is essential that you learn where they are and how to work them.

It is impossible to exaggerate the importance of being prepared, both practically and psychologically, for the possibility of trouble at sea. Such preparation enables skipper and crew to swing immediately into efficient action the moment the merest beginnings of a problem appears, and the potentially serious emergency is nipped in the bud. If, however, the crew has to run round in circles wondering what to do, then trouble will snowball, often with incredible rapidity, until it is completely beyond the ability of the crew to handle. Very often people then start dying and boats sink.

Chapter 7
Handling Sails

Some modern yachts are so efficiently rigged, with sails that can be simply rolled up for stowing or reducing in size, that most of the time there will actually be little or no sail handling to do. On more conventional yachts, however, there will be a great deal to do and it is most important that you do it all correctly. Unless controlled properly a sail can be a very dangerous thing, especially when it thrashes about in strong winds.

Photo 29 This small yacht has a roller jib, rolled up tightly while she is motoring. When the furling line is released, the jib is set simply by hauling on the leeward sheet.

Roller-furling Sails

In Photo 29 you can see a modern yacht's roller-furling jib. This type of jib can be rolled up round its stay, when not required, by simply pulling on a rope that is wound on a drum at the foot of the stay. As the rope winds off the drum so the complete structure rolls round, thus winding the sail round the stay. A roller-reefing jib works in exactly the same way, but the sail is specially cut so that it can be set part-furled, to produce a smaller sail for use in strong winds, which cannot be done with a roller-furling jib.

Mainsails can also be fitted with roller-reefing gear, enabling them to be wound up inside the mast, or round a pole abaft the mast. They are not very common as they do not provide such a huge handling advantage over conventional sails as do the roller jibs.

Conventional Sails

These sails are rigged quite differently to the roller types and have to be set and *HANDED* (taken down) whenever required. Mainsails are reefed to make them smaller in strong winds (just like a dinghy sail - see Chapter 3), and headsails changed for smaller ones. Clearly a great deal more knowledge and skill are required to handle these sails, but they do have the advantage of being simpler and more reliable.

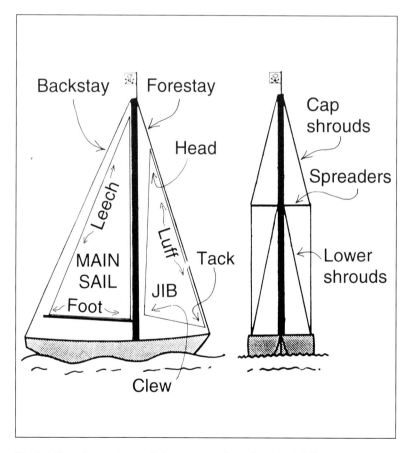

Fig 37 The shrouds and stays are wires that hold the mast up. The three corners of the sail - tack, head and clew - apply to both sails, as do luff, leech and foot.

Photo 30 A small yacht sailing to windward with mainsail and a conventional jib set. Note how the jib is trimmed so as to direct the flow of wind around the leeward side of the mainsail. The wind is coming from the yacht's starboard side, and she is said to be on starboard tack. The trimming of this 'slot' between jib and mainsail is critical for good sailing performance. If the mainsail 'backwinds', the jib is too tight; the boat will sail inefficiently.

Where a roller jib is permanently attached to the forestay and rolls around it, the conventional one is attached as required, then pulled up it by a halyard. The traditional attachment is a series of *HANKS* (special clips) fastened to the *LUFF* (front edge) of the sail at intervals, which can be clipped over the stay. The *TACK* of the sail is attached to the stemhead fitting, the *CLEW* to the sheets and the *HEAD* to the halyard. See Figure 37. The halyard leads to the top of the mast and back down to a cleat. The sail is hauled up as tight as possible, often by taking the halyard round a winch, then the halyard is cleated off. See Photo 30.

Some racing boats may be found to have a *HEADFOIL* in place of the conventional wire forestay. This is like a grooved aluminium rod and the rope in the luff of the jib is simply fed up inside the groove. This improves the flow of air over the jib, and thus its efficiency, but does mean the jib is not held captive by its hanks round the stay when it is dropped onto the deck.

In practice, especially at sea, the technique is as follows:

1   Pull sailbag on deck; tie bottom to nearby guard-rail.

2   Prepare halyard and sheets; tie to PULPIT ready for attaching.

3   Sit in pulpit; pull out tack and shackle onto stemhead.

4   Pull out sail steadily, hanking onto forestay as you go.

5   Attach sheets as clew comes out of bag.

6   Attach halyard as head comes out of bag; hoist sail; sheet in.

See Photo 31.

Photo 31 The tack of a conventional jib. The small blob near the bottom of the luff is a hank and the railing round the bow is the pulpit. It should be apparent that the jib can be controlled well if you are firmly wedged in here and looking aft. To dowse the sail you can gather the foot towards you until the whole sail is firmly held within your arms.

Note the halyard winches on each side of the mast, and the method of stowing the coils of the halyard. Instead of tucking the coil behind the halyard (see photo 25), pull the short length of rope between cleat and coil through the coil from the mast side, then twist the bight and hook it over the cleat so that the loop formed holds the coil close to the cleat. You may have to adjust the length of the bight to hold the coil snugly.

Photo 32 The mainsail is seen quite clearly here, together with its sheet (from the end of the boom to the stern) and topping lift (from the end of the boom to the top of the mast). A stowed sail is shown in Photo 29. The jib is goose-winged, and you can see both sheets leading from the clew to the sides of the boat. See section on Special Techniques at the end of the chapter. The canvas sheets on either side of the cockpit with the name on are called *DODGERS*; they give some protection from spray and wind to the people in the cockpit. See Photo 30 for a better view.

The secret of efficiency here lies in the correct bagging of the sail when it is handed. Always stuff the sail in loosely, putting the head in first, then the body of the sail, then the clew, then the tack last. If jibs are always stowed like this then they always come out as described above, and you can secure the tack first then steadily hank the sail on. At all times the sail is either attached securely to the boat or safely in the bag, and thus under control.

The mainsail is rather easier to set as it is normally left on the mast and boom all the time. All that is required basically is to pull it up with the halyard, which is probably also left attached. See Photo 32. In practice the technique is as follows:

1   Take weight of boom on *TOPPING LIFT*.

2   Slacken mainsheet so boom can swing freely in the wind.

3   Remove sail tiers.

4   Haul on halyard until sail is right up; cleat halyard.

5   Heave down on *TACK TACKLE* or tension halyard to tighten luff.

6   Slacken topping lift so sail holds boom up.

Trimming Sails

Trimming sails on the yacht is no different in principle from doing the same on the dinghy, but the gear is likely to be more complicated. There will almost certainly be more than one sail, and all will be large enough to require mechanical assistance for heaving them in - probably powerful winches to wind in the sheets.

Handling these large sails clearly requires a great deal more skill and care than does handling the little one on a dinghy. In fresh winds the strain on a jib sheet can be considerable, and the attendant dangers very real. Thrashing sheets and shackles can be dangerous, and if an unwary hand is caught between the turns of a sheet round a winch and the barrel of the winch then fingers can be crushed by the forces on the sheet. You must remain constantly alert to these dangers while moving around a yacht under sail, and especially when you go for'ard to set or hand jibs.

When trimming pairs of sails together - mainsail and jib - you should trim the jib to suit the wind, then trim the mainsail till it just 'goes to sleep' in the wind that is being steered round it by the jib. If the sails are trimmed correctly the boat should steer easily, without too much weight on the tiller. If the mainsail is pulled in too tight, especially on a reach, the boat will develop excessive '*WEATHER HELM*', meaning she is constantly trying to swing up to the wind. This can be very tiring and make accurate steering difficult. If it cannot be cured by trimming the sails properly, the mainsail will need reefing, or the jib increased in size, in order to restore the correct balance between them. Report such difficult steering to the skipper immediately. See Chapter 8.

Working the Winches

The winch in Photo 33 is worked by simply wrapping three turns of the sheet round it and winching back and forth on the handle while hauling on the sheet. Although designs vary (some have handles on the top that are wound round and round; some have two speeds; some are *SELF-TAILING* in that they grip the sheet as you wind so that you do not have to haul on it) the principle remains the same. In strong winds winches need to be used with care as they can easily trap fingers with considerable force, and if turns of the sheet are allowed to ride over one another the whole lot can jam solid, making it impossible to release the sheet. The dangers in this should be apparent.

These riding turns are most likely to happen when heaving in jib-sheets too quickly, without paying attention to the simple techniques described in the next paragraph. If you get riding turns jammed on a winch, a line should be rolling hitched onto the sheet ahead of the winch, passed round the jammed winch and across the cockpit, to be hauled tight on the other winch. With the weight of the sheet removed from the riding turns they can usually be cleared fairly easily, replaced cleanly on

Photo 33 When you are winching in a sheet under strain you should always have the end tucked round a cleat, then if the strain suddenly becomes too much to hold you can quickly add more turns to the cleat to hold it. Note how the crew stands back from the handle and braces her legs against the side of the cockpit; this enables her to get a really good heave on the handle without overbalancing. Handles normally wind clockwise, but sometimes a lower gear is obtained by winding back the other way.

the winch and the weight taken again before releasing the rolling hitch. A line like this, used to take the weight partway along a rope, warp, anchor chain or whatever, enabling you to shift the lead or the securing point, or clear tangles, is known generally as a *STOPPER*. It is an extremely useful gadget.

The general principle of working a winch is to put just one or two turns on at first (check which way round the winch rotates before putting them on!) and haul in as fast as possible until the weight of wind begins to come on the sheet. Then carefully wind on further turns to a total of three or four (depending on wind strength), keeping firm hold of the weight while doing so and your hand well clear of the winch in case you get suddenly pulled towards it. Make quite certain the turns are cleanly on the barrel and not riding over one another, then wind in on the handle while leaning back on the sheet so that your weight helps to hold it and the chance of being overbalanced is reduced. Then turn the sheet up round the cleat perhaps four or five times to ensure that it grips securely. This is much easier and safer with two people - one winching and the other tailing (hauling on the sheet). See also Chapter 6 for more information on winches and handling sheets.

Jib sheets come through fairleads to make sure they run onto the winches at a suitable angle as well as pulling evenly on the sail. To prevent the sheets flying out through the fairleads when they are let go they should have Figure of Eight knots tied loosely in their ends. See Figure 38.

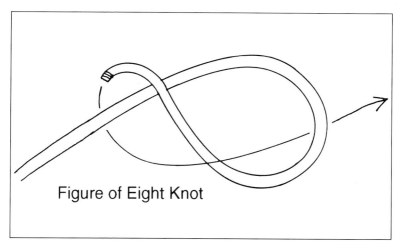

Figure of Eight Knot

Fig 38 Make this knot at least a foot away from the end of the sheet so that if it is pulled tightly into the fairlead there is enough tail to get hold of to pull it out again.

Reefing Sails

When the wind becomes too strong to sail safely with the full-size sails, they must be reduced in size. Roller-reefing jibs are rolled up the required amount; conventional ones handed and replaced with smaller ones. Mainsails are reefed in three main ways: rolled into mast; rolled round boom; lashed to boom. The first is performed simply by letting go the OUTHAUL and hauling on the furling line, much as a roller jib is reefed by letting go the sheet and hauling in the furling line.

The second system is generally referred to as roller-reefing, and is usually carried out by winding a handle at the for'ard end of the boom to rotate it, at the same time slackening off the halyard. The sail is wound round the boom. The weight of the boom must be taken on the topping lift. See Photo 34.

Slab reefing, as the last system is called, has the great merit of simplicity as it requires no complicated worm gears and handles. The weight of the boom is taken on the topping lift and the sail lowered until the REEF CRINGLE at the luff can be hooked into the hook on the boom. The reef tackle is then hauled on to pull the leech reef cringle tightly down to the boom, so the sail is secured at each end. The loose sail between is then tied up with the REEF PENDANTS using reef knots (see Chapter 6). See Figure 39.

Photo 34 ⇒

The crew winds the sail round the boom while easing away the halyard round its winch. Note the sliders on the luff of the mainsail that slot into a groove on the mast to hold the sail against the mast. There is usually a pin in the bottom of the groove to stop the sliders falling out; this needs to be removed when reefing, so that they can fall out. It can be difficult to keep a good shape in the mainsail after it is roller-reefed, so care must be taken as the sail must be efficient in strong winds. Top the boom slightly and try to pull the leech out along the boom while rolling. Practice with a friend to see how to get the best sail shape when reefed.

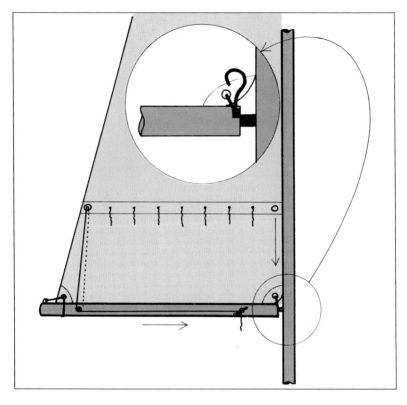

Fig 39 To reef this type of sail you must take the weight of the boom and sail on the topping lift, then lower the halyard until the luff *REEF CRINGLE* (eye) can be fitted onto the hook at the tack of the boom. Then haul on the *REEF PENDANT* through the leech cringle until the clew of the sail is pulled down tight to the boom. Then reset the halyard tightly and slacken the topping lift.

Racing boats reduce sail while roaring along at full speed, and it can be very wet sat in the pulpit changing to a small jib while crashing to windward. Civilised cruising sailors, on the other hand, stop the boat by *HEAVING-TO* so that she sits quietly and calmly, providing a safe, dry and comfortable platform for reefing, or doing any other tasks on deck if necessary. The simplest form of heaving-to most likely to be used for this consists merely of *BACKING* the jib by hauling it tightly in with the weather sheet. This can be accomplished by simply tacking the yacht without changing the jib to the other side. See Figure 40.

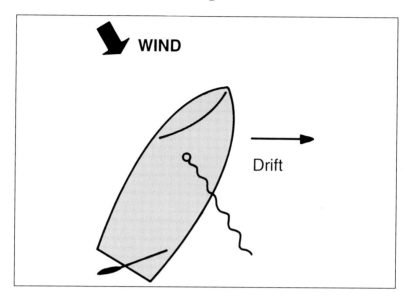

Fig 40 Hove-to on port tack. Note the backed jib, the flapping mainsail and the tiller lashed to leeward. The rudder makes the boat swing towards the wind, whereupon the backed jib pushes her head away, and after a while the boat settles into a state of equilibrium, drifting sideways as shown.

Handing and Stowing Sails

Removing the sails (handing sails as it is known) is simply the reverse of setting them. Take the weight of the boom on the topping lift then lower away on the halyard. Haul the boom inboard with the sheet as soon as the wind permits, and drop the sail right down. The mainsail is normally left on mast and boom and lashed securely to the boom using short ties of rope or elastic. The sail should be hauled aft along the boom to spread it evenly, then furled as tightly and lashed as securely as possible to prevent the wind blowing it all apart.

Handing the jib is slightly different as it is normally removed completely and stowed below in a bag. As mentioned above, the jib should be stowed in the bag head first and tack last. This makes both handing and setting easier as the sail is kept under control throughout both processes, being either hanked on steadily up from the tack or unhanked down from the head. The sail is thus either in the bag or hanked on, and so under control. This is most easily done by sitting firmly wedged in the pulpit with the bag tied to the guard-rail nearby.

A temporary stow is often made just by lashing the sail along the guard-rail as one would the mainsail along the boom. Halyard, sheets and hanks remain attached so that the jib can be very quickly set again when required. Be very careful when working with sheets and halyards not to let the loose ends trail over the side where they could tangle in the propeller. This frequently happens when sails are dropped in too much of a rush. When handing a jib the sheet can be left tight until the sail is almost down then released when little wind remains in the sail to flog it over the side. You must then haul the foot of the sail forward to pull the clew and sheets on to the foredeck out of the way.

In Photo 35 you can see a small cruiser on her mooring with sails down and everything neatly and firmly stowed. It is quite incredible the power of a strong wind to pull apart furled sails, so this neatness has a purpose other than mere smart appearance. You can see how tightly the mainsail is furled and lashed. The boom is held securely on the topping lift and prevented from flying about by hauling down tightly on the mainsheet. The jib sheets are fastened to the forestay, and hauled tight onto their cleats, and, if you look carefully, you can see short lines tied from the shrouds (wires from masthead to each side of the boat) to the halyards to pull them clear of the mast. They are known as *FRAPPING LINES* and without them the halyards will bang and rattle incessantly against the mast. Lengths of elastic shock cord with hooks on the end are often used. Furling jibs theoretically need only rolling up with the furling line. In practice they can come unfurled in strong winds and flog

Photo 35 A neatly and tightly stowed mainsail that should not blow to shreds when it is windy on the mooring. The jib sheets are secured to the halyard and hauled partway up the forestay to keep them clear of the deck. The white blob here is a small buoy attached to the mooring rope that makes it easier to pick the rope up as the end stays afloat. See Chapter 9. It is tied to the forestay at jib sheet height to signify that the boat is moored.

themselves to shreds, so a secure lashing should be put round them by the clew. They should also be rigged so that the furling line will roll the sheets right round the sail two or three times to provide extra security.

The FALLS of the halyards should be coiled and stowed as described in Chapter 6. When the halyards are required - for setting or handing the sails - the coils should be released from their stowages and dropped onto the deck so that the parts leading up to the cleats lie on top. This is particularly important when handing sails as it allows the halyard to run out from the top of the coil as the sail is lowered, without tangling. Drop the coils somewhere secure so they will not fall or get washed over the side, especially if the engine is running.

Special Sails

Various types of large, lightweight sails are often set in light weather in order to give extra speed, and some of them require specialised handling. The *SPINNAKER* - a very large parachute-like sail - can be set downwind, flying from the masthead rather than hanked to the forestay, but if you are cruising you may well come across some sort of cross between one of these and a jib, as they are much easier to handle. See photo 36.

Storm sails are special, very strong sails designed for use in storm conditions. Modern yachts sailing inshore do not generally use them as their sails are strong enough to stand up to the conditions they are likely to meet. Yachts sailing offshore, however, very probably will have them as they would expect to encounter much worse weather than the coastal sailing boat. Old-fashioned yachts with sails made of traditional materials that are not so strong as modern ones may well have storm sails for use even inshore.

The *STORM TRYSAIL* is a small triangular sail that is set in place of the mainsail. Its purpose is to produce as long a luff as possible for sailing efficiently to windward, together with a sheeting system that enables it to be set without the boom (which can be a danger in rough weather as it swings about). It is rigged in a quite different way to the mainsail and it is most important that you know precisely how to do it, and practise in harbour, so that you know what to do when it is blowing hard in the middle of the night and the skipper decides to set it.

The *STORM JIB* may well be set on a boat carrying a roller-furling or roller-reefing jib, and will be hanked onto a small stay that can be set up inboard of the jib when required. This is simpler, stronger and more efficient than rolling up a large jib to a very small size. On a yacht setting conventional headsails the storm jib will be simply the smallest and strongest jib on board, hanked on the forestay just as the others are.

Photo 36 This multi-coloured sail is called a *CRUISING CHUTE*. It is similar to a spinnaker but flatter and smaller and much easier to handle. Spinnakers have both corners flying out at the end of a sheet, but this sail is tacked down at the stem and controlled by one sheet on the leeward side. The dark object at the top of the sail is a long narrow bag that can be pulled down over it to spill the wind and turn it into a dead sausage that can be easily and safely hoisted and lowered by a small family crew. Note the roller jib and the spinnaker pole on the front side of the mast near the bottom. This can be used to goose-wing the jib.

Special Techniques

There are two problems associated with running dead before the wind and you will almost certainly be called upon to take measures to deal with them. The first is that the mainsail will blanket the jib so that it does not fill with wind. This is resolved by setting the jib on the opposite side to the mainsail, a technique known as *GOOSE-WINGING*. You may be provided with a short pole that fits between the mast and the clew of the sail to hold the latter out and stop it flapping.

The other problem is the risk of an accidental gybe if the wind shifts suddenly or the helmsman wanders off course. Such a gybe in strong winds can break the boom and also the head of anyone in the way, so it can be extremely dangerous. It can be prevented by rigging a long line very tightly from the end of the boom to the bow of the boat. This is called a *PREVENTER*, as it will prevent the boom crashing across the boat if she gybes. The best way to rig it is to slacken the mainsheet off a bit too far then tighten the preventer, then haul in on the mainsheet so that the preventer is bar-taut. The preventer should have enough surplus line to surge round the cleat and allow the boom to be gradually eased across the boat under control if it does gybe.

Chapter 8
Steering and Watchkeeping

Although the ultimate responsibility for the safety of a yacht lies with the skipper, it is not practical for him to steer and keep watch all the time. He needs to navigate, plan and prepare passages, ponder on the weather, repair broken gear, sleep, attend to his ablutions and so on. When not actually in the *COCKPIT* (steering well) he must delegate someone else to take charge of the yacht, and that someone is called a watchkeeper.

On a well-run yacht there will be a proper watchkeeping system that makes provision for all the crew to take turns on watch. The system used will depend on the number of crew and their experience, and the conditions prevailing.

Unless a boat is sailing very short-handed, a skipper should never keep a regular watch. This will ensure that he is always available, refreshed and rested, when required to attend to any matter that the watchkeepers cannot handle. He will not be able to do this if he spends large parts of the day and night trying to keep awake at the tiller.

So a lot of responsibility rests with the watchkeepers and it is essential that the skipper be able to rely on them completely. By them, of course, I mean you. And the most important thing you must realise is that you are not there to take charge of the boat, but simply to keep her safe in the absence of the skipper. It is not your place to make decisions about *anything* unless you have been specifically instructed to do so. If a skipper can trust a complete novice to do exactly as he is told, then he can usefully employ that person to help him sail the yacht. Even if the watchkeeper calls him every three minutes in fear and trembling, it will still enable the skipper to get some rest or carry out an important task that he would not otherwise be able to do. If, however, the watchkeeper thinks he knows it all and cannot be trusted to do as he is told, then he is best left ashore with a video game, in which people's lives are not at risk. A watchkeeper who can be trusted, however inexperienced, is far more use to a skipper than a knowledgeable one who cannot.

Responsibilities of the Watchkeeper

In short - the safety of the yacht in the absence of the skipper. And this involves rather more than simply 'being there' and steering in the right direction. Steering, in fact, is probably the least important of the watchkeeper's tasks, and is usually delegated anyway if the watch consists of more than one person. These days it is very likely in fact to be delegated to an *AUTOPILOT*, leaving the watch-keeper to concentrate on keeping watch.

And for a watchkeeper the essential task is to watch: the movement of other shipping that might come close enough for there to be a risk of collision; the course steered by his own boat to ensure it is that ordered by the skipper; the weather for any sign of change and deterioration; the sails and rigging for correct trim, onset of chafe or breakdown etc; the engine, if running, and attendant gauges for signs of trouble; the bilges for the unexpected arrival of water; other persons in the watch for seasickness, tiredness and lack of concentration that could lead to poor steering or inefficiency if trouble arises; and so on.

At its simplest level watchkeeping consists of this plus an instruction to call the skipper the moment anything gives rise to doubt. As the level of experience of a watch-keeper rises so can his responsibilities. The list of skipper's instructions can also get simpler. The Mate of the vessel, for example, can probably be left virtually to his own devices with no more detailed instructions than 'have a quiet watch'. The young, inexperienced chap can be told to call the skipper on sighting anything, and the average watchkeeper can be given a list of instructions stating clearly the occasions on which the skipper should be called.

This list will often be kept in a *Skipper's Night Order Book* left on the chart table. In the back will be noted the standard events that require the calling of the skipper - vessel on collision course; increase in strength or change in direction of wind, and so on. In the front can be listed specific instructions for each watchkeeper on a particular night, such as - 'Geoff: call me on sighting the Longships lighthouse; should be fine on port bow at about 0330. If not seen, call me at 0345'. You can see a sample page in Figure 41.

As well as keeping watch, the watchkeeper must also enter up the *DECK LOG* hourly with all the information required by the skipper. This usually entails noting wind speed and direction, sea state, *BAROMETER* reading (atmospheric pressure), course steered, estimated *LEEWAY* (sideways drift caused by wind), course alterations and so on. If the engine is running, the readings from all gauges should also be noted, although this is better done in a separate engine logbook, which can also contain details of maintenance etc. See Appendix 2.

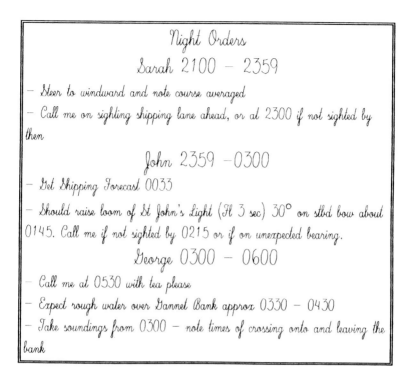

Night Orders

Sarah 2100 – 2359

– Steer to windward and note course averaged
– Call me on sighting shipping lane ahead, or at 2300 if not sighted by then

John 2359 –0300

– Get Shipping Forecast 0033
– Should raise loom of St John's Light (Fl 3 sec) 30° on stbd bow about 0145. Call me if not sighted by 0215 or if on unexpected bearing.

George 0300 – 0600

– Call me at 0530 with tea please
– Expect rough water over Gannet Bank approx 0330 – 0430
– Take soundings from 0300 – note times of crossing onto and leaving the bank

Fig 41 Sample night orders from a skipper's Night Order Book.

Watchkeeping Routines

There are various ways in which watches can be arranged, depending on circumstances. The traditional naval four-hour watch is too long generally for a small boat and a basic three hour watch is better, giving time enough off watch for proper rest (with three on and six off) together with a short enough watch to remain awake. This basic structure can be varied to suit: six hour day watches in fine weather being an excellent way of getting a good sleep when short-handed and working watch and watch about; and two hour night watches being quite long enough when alone in bad weather at night. In Figure 42 there are some examples of watchkeeping rosters for various combinations of crew size and experience, and conditions encountered.

Watchkeeping Systems

Two watches

Night	- 3 hours on
	- 3 hours off
Day	- 6 hours on
	- 6 hours off

Three watches

3 hours on
6 hours off

Four watches

3 hours standby (domestic and maintenance chores)
3 hours on
6 hours off

Fig 42 Some typical watchkeeping routines used aboard yachts on passage.

Whatever system your skipper works, and there is no legal requirement to stick rigidly to the same one at all times, there are certain routines that should be adhered to. The new watch should be called fifteen minutes before they are due to take over, which gives them time to dress, switch brains into gear, collect their night vision, and turn up bearing cups of tea five minutes before the hour. They can then acclimatise to the conditions and have a proper handover. This handover is important. The off-going watchkeeper should ensure that the relieving one is given all information about shipping, recent weather forecasts or changes, current skipper's instructions, sails set, the boat's position on the *CHART* (map of the sea), any visible land or sea marks, the course being steered and so on. He should then check that the deck log is up to date before disappearing below.

Steering on Watch

If there is more than one person in the watch this should be overseen by the watchkeeper. Half hour tricks are usually long enough, particularly at night or in bad weather,

Photo 37 This is a steering compass fitted on the fore side of a cockpit where it can easily be seen by the helmsman under all conditions. The course steered is indicated by the white line at the forward end.

as staring at a *COMPASS* (a gadget that shows the direction you are steering) is very tiring under such conditions. It is much better to steer on a landmark or star and glance at the compass now and then to ensure that the course is correct. Bear in mind, however, that stars move and tidal streams can cause landmarks to, in effect, do the same. See Photo 37.

He must also check to see what course is actually being averaged, as opposed to the one ordered. This *HELMSMAN'S ERROR* is quite normal and acceptable, but it is most important that it be noted for the benefit of the navigator. Different helmsmen will have different errors and this must be watched for. The deck log should have separate columns for course ordered and course averaged, as well as estimated leeway. With all this information regularly assessed, the navigator will be able to estimate much more accurately the yacht's position should he have no electronic navigator to tell him.

The course averaged when sailing to windward is especially prone to variation, due to the need to continually swing the boat slightly into wind in order to ensure she is sailing as close to it as possible. This effect is accentuated in rough weather as the helmsman needs to *LUFF UP* a bit (swing towards the wind) in order to climb the face of each big wave, then *BEAR AWAY* (away from the wind) to most efficiently sail down the back of it.

Steering a yacht with a tiller is no different in principle from steering the dinghy, although the weather helm will be heavier. Steering with a wheel, however, is rather different. Instead of sitting to weather of a tiller and steering by pulling gently and steadily against the weather helm, you generally have to sit behind the wheel and wind it back and forth rather like the wheel of a car. It is not so easy to feel the weather helm and general flow of water past the boat through a wheel as it is to feel it through a tiller, but with a large boat a wheel tends to be less tiring. Having learnt that a tiller steers the stern of a boat, you need to adjust to thinking of a wheel as steering the bow. In other words, putting the tiller to port moves the stern to port and thus the boat turns to starboard. As in a car, however, winding the wheel to port turns the boat to port.

The balance of the sails has a most important effect on steering, and the watchkeeper should keep them trimmed so as to ease steering as much as possible. This was discussed in the last chapter. If it gets too difficult then he must either change the sailplan or call the skipper, depending on his instructions. Reaching fast down big seas generally produces the heaviest steering conditions, due to a tendency to broach, and effort should be made to concentrate sail area forward so that the sails tend to pull the boat as straight as possible. Inexperienced helmsmen may have to stand down in such conditions, as not only will it be extremely difficult for them to average a reasonable course, but also they may allow the boat to broach, which could have serious consequences. See Figure 17 in Chapter 4.

Surviving a Watch

A three hour night watch in wet, cold and windy weather can be very wearing indeed, but there are certain techniques which can be employed to make it more bearable, not just for the comfort of the watchkeepers as such, but because the more miserable a man is the less efficiently he will steer and keep look-out. It is the responsibility of the watchkeeper to ensure that everyone in his watch is kept warm and efficient.

Proper sea-going clothing that will keep you warm and dry is obviously important, but simply sitting still in it for long periods reduces both its efficiency and your own. Creeping lethargy brings on a mental depression and a slowing of the bloodstream, which then generate tiredness and coldness; so it is important to move about frequently, stamping the feet and swinging the arms so as to send the blood hurtling hotly through the veins. Foul weather gear does not create heat, it simply helps to keep in that which you generate. If you sit crouched and miserable in the corner of the cockpit, your body will not make heat for the clothing to keep in. Under the oilskins you will find that many thin layers are more effective insulation than a few thick ones, as they create a multitude of air pockets.

Morale is boosted considerably by talking, brewing up, pottering about trimming sails or repairing things (weather permitting) and so on. If more than one man is on watch and conditions are reasonable then you can beneficially take turns to sit below in the warm for ten minutes. A useful occupation at quiet moments is to run through in your mind the emergency drills - man over-board especially - so that if one suddenly arises you will have the routine fresh in your head. See Chapter 6.

Keeping a Look-out

This is a vitally important part of watchkeeping and should be approached in a thoroughly professional manner. The basic technique should be to sweep slowly round the whole horizon, from right ahead to right astern, with the eyes; then do the same with binoculars. Pause to relax the eyes for a few minutes then repeat the whole process. See Photo 38. If you think you caught a glimpse of something but cannot see it when you look hard, then look slightly to one side as the corner of the eye is more efficient at spotting such things than the middle. This is particularly effective when searching for faint lights at night.

Photo 38 Keeping a look-out with binoculars. Hold them as steady as you can and sweep slowly and steadily round the horizon so that you do not miss anything. Ahead of the crew is another type of compass.

You must ensure that you can see clearly all round the yacht. If there are blind spots caused by sails (low-cut jibs are often culprits) or deckhouses, dinghies etc, make quite certain, by moving position if necessary, that you look round them each time you survey the full horizon. This is especially important at night when faint lights may bob up intermittently in waves.

You may be told to look out for a light on a buoy or lighthouse as the yacht approaches land. These navigation lights flash with different characteristics (eg 3 red flashes every fifteen seconds, or one white every five seconds, etc), which enables a navigator to identify them. The timings must be checked carefully three or four times using a watch, as waves can often obscure flashes from a buoy and make the characteristic seem wrong.

In fog you must also look with your ears. The sound of a ship's engine can be heard much further under water than through the air, so try going below now and then and listening with your ear pressed to the hull underwater. It will be very difficult to gauge the direction, but this will at least give some advance warning and an indication of whether the sound is approaching you or moving away. Look-outs should be posted for'ard so as to see that bit further, and also to listen away from noise in the cockpit or the engine. It is extremely difficult to assess visibility at night unless you can see lights gradually fade, so if in the slightest doubt about the visibility you must call the skipper. There should be a *FOGHORN* on board, that you can use to alert other ships to your presence, and the skipper will doubtless instruct you in its use.

Assessing Risk of Collision

If you see another vessel that appears to be closing with you, you must check its movement by taking a bearing of it with the *HAND-BEARING COMPASS* (a small compass that can be held in the hand and aimed at an object to check its direction). A few minutes later take another bearing and see how they differ, if they do. If the bearing remains the same then the vessel is on course to collide with you; if it moves towards your bow the vessel will pass ahead of you, and if it moves towards the stern it will pass astern of you. If the bearing is steady or changing only slowly then the skipper or watch leader must be alerted immediately as a fast modern ship can reach you from coming over the horizon in only a few minutes. See Figure 43.

At night the positioning of a ship's lights can be of great assistance in determining its movements. In Figure 44 you can see the lights carried by most merchant ships. It should be apparent from a study of the diagram that the two steaming lights show which way the ship is pointing (lower one is forward), and that any change in their distance apart will provide immediate indication that the ship is altering course.

Many ships, especially ferries and fishing boats, will have so many deck and cabin lights that it will be quite impossible to see the navigation lights in amongst them.

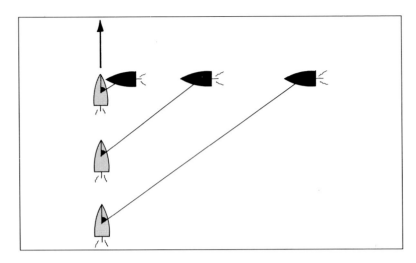

Fig 43 It should be apparent that if the relative bearing of the black vessel from the white one remains constant the two will collide.

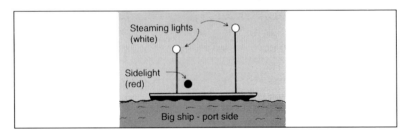

Fig 44 Sidelights (red to port and green to starboard) are much dimmer than the steaming lights. If you can see them you are too close for safety, so virtually all collision avoidance at sea must be done using the disposition of the steaming lights. Different types of vessel show different collections of lights. As you become more proficient at watchkeeping you will need to learn more of them in detail.

It takes some experience to learn to assess the lights and behaviour of a ship at night, and speeds and distances can also be extremely deceptive. If in the slightest doubt about lights you can see, call the skipper immediately and tell him it is urgent.

Keeping the Navigational Plot

Navigation is not the job of the watchkeeper but you may find yourself expected to plot your position on the chart each hour, having taken it from an electronic navigator. This will enable the skipper to briefly glance at the chart and see your position each time he comes on deck. In an emergency with the skipper incapacitated you must be able to work out your position on the chart and transmit it in a radio distress call. See Appendix 3.

The electronic navigator will give the yacht's position in terms of what are known as *LATITUDE* and *LONGITUDE*. Latitude is your distance north or south of the equator and longitude the distance east or west of the Greenwich Meridian (a line running north-south through Greenwich in London. See Figure 45. These distances are noted as degrees and minutes, measured along the circumference of the circle that is the Earth, and they are plotted using the scales at top and side of the chart. See Figure 46. To plot a position lay the parallel ruler along the required latitude and measure along it from the side of the chart the longitude, using a pair of dividers. Mark the position with a small cross in pencil and write both the time and the Log reading alongside it. See Figure 47.

If the radio navigator is working then you can, of course, simply transmit in a distress message the position shown on it, without having to plot it on the chart. If, however, it has broken down, you will appreciate the value of regularly plotting the position on the chart every hour. You must then work out where you have travelled since the last plotted fix, mark that on the chart and read off its latitude and longitude. The ability to do this could save the yacht and all aboard her if everyone happens to be struck down apart from you. See Figure 48. Do bear in mind, however, that those searching for you will be infinitely better navigators than you are, so if in any doubt at all about your ability to plot your current position accurately (the yacht may have drifted out of control for a while if she is in serious trouble), then give the last plotted position and its time and describe what the yacht has been doing since.

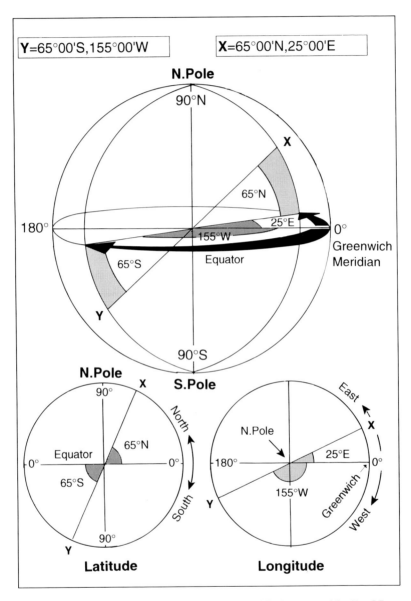

Fig 45 The position of X is described as 65 degrees North, 25 degrees East. Y, diametrically opposite, is 65 degrees South, 155 degrees West. Careful study of the diagrams should make this clear.

118

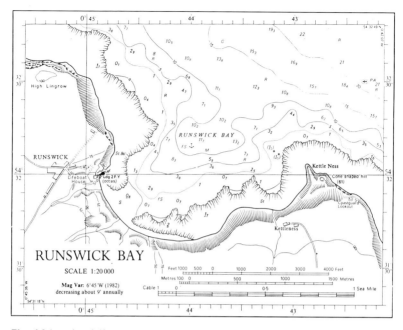

Fig 46 Look at the scales around the edge of the chart. Along the top and bottom run the Longitude scale; it increases from right to left so is West of Greenwich. The Latitude scale at the sides increases upwards so is North of the Equator. Kettle Ness is approximately 54 degrees 32 minutes North and 0 degrees 42.9 minutes West.

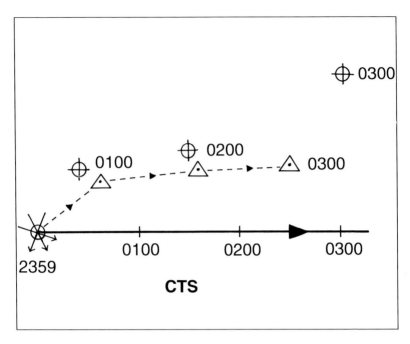

Fig 47 This plot shows various types of position that a navigator may mark on the chart. On the left is a *FIX* consisting of compass bearings of three objects such as beacons, lighthouses, trees etc, taken at midnight.

The line with the arrow extending from this fix indicates the course that the yacht is steering along (CTS=Course To Steer), and the small line crossing it at hourly intervals show where the yacht should be after each hour sailing along that course, if she is not carried off it by tide or wind.

The triangles above this line are the *ESTIMATED POSITIONS* for the same times, being where the navigator estimates the yacht to actually be, taking into account tide and wind influences.

Above these are Radio Navigator Fixes. A prudent navigator will always run a plot like this, enabling him to constantly check his EPs (Estimated Positions) against his fixes. Any major discrepancy clearly indicates an error, as at 0300.

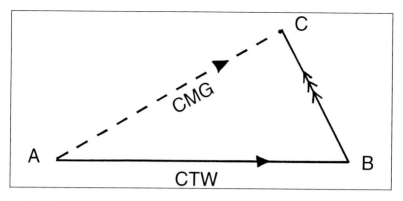

Fig 48 From a known position at A, plot the position of B, estimated from the course you have sailed and distance run through the water (CTW = Course Through Water). Mark the line B - C consisting of the distance and direction you estimate tide and/or wind to have set you during this same time. C is your actual position and A - C your Course Made Good (CMG) over the ground - ie on the chart. Measure the latitude and longitude of this position using parallel rulers and dividers.

Finally, let me reiterate what was said at the beginning of this chapter about watchkeeping, as it is so very important. Remember always that you are simply keeping watch for the skipper; you are not running the ship. If in the slightest doubt about anything, call the captain. That is what he is there for.

Chapter 9
Mooring and Anchoring

After the taste of independence that you had while handling the dinghy on your own, you will find your part in these manoeuvres very much a matter of being part of the team. However well the skipper handles the boat, all will come to nought if you - the crew - get things wrong on deck. So you carry a great deal of responsibility when the boat is being brought alongside or anchored.

At the same time, however, you must understand that the final overall responsibility for the boat lies with the skipper, and with him alone. Your responsibility, in effect, is to do exactly as he tells you; so that he can then be certain of being able to rely on you. If the skipper knows that he can depend on you to carry out his orders to the letter, promptly and efficiently, it will give him the confidence to pull off difficult manoeuvres that may at times be necessary to get you all out of trouble.

Mooring Alongside

There are two aspects to bringing a boat alongside that the crew must keep in mind at all times. One is tying up and fendering the boat on arrival, and the other is the possibility of having to use fenders and warps to help manoeuvre her while she is getting alongside. Let us look at the first one.

In Figure 49 you can see the four basic ropes that hold a boat alongside a quay or marina pontoon. Each one has a specific task to perform, and it is important that you understand what that is if the skipper is to be able to rely on you to set them up properly. Rarely in real life will it be possible to lead these warps as neatly and logically as they are shown in the drawing, but if you appreciate what each one is supposed to do you will be able to rig them well enough, however awkward the situation.

Basically, we can say that the *HEAD* and *STERN* ropes hold each end of the boat into the berth, while the *SPRINGS* stop her from moving back and forth along the berth. To allow for the boat rocking in waves the head and stern ropes should be left a little slack, and should lead out along the jetty a bit rather than directly to shore. Because the springs are long and almost in line with the boat, they do not need to be slack as they will simply rise up and down as the boat rolls. They should, in fact, be kept tight or the boat will rub back and forth along the fenders.

Fenders should be evenly spaced along the hull and hung at the correct height for the jetty, tied to the *GUARD-RAILS* (safety wires round boat) with Clove

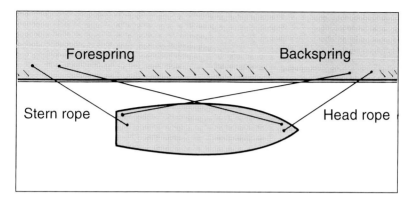

Fig 49 A yacht moored properly alongside a jetty or pontoon.

Hitches (see Chapter 6) so that they can be rapidly adjusted for height and position should it be necessary. When you are secured in the berth, however, you should then tie them to the bases of *STANCHIONS* (posts supporting safety wires; see Figure 50) with the Round Turn and Two Half Hitches (see Chapter 5) as this knot is more secure than the Clove Hitch, and the stanchion bases stronger than the guard-rails. Keep one fender free, lying on deck ready for use anywhere in emergency. Fenders should have *LANYARDS* (short lines) on both ends so they can be hung horizontally to span vertical piling on a jetty.

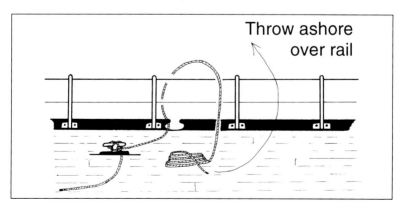

Fig 50 The correct way to prepare a mooring warp for berthing.

The mooring warps should be prepared as shown in Figure 50 so that the coils can be simply lifted from the deck and taken ashore, without having to lead them under the guard-rails. This way is quicker and less prone to mistakes in the heat of the moment, as the warp will automatically lead correctly on being hauled tight and secured.

There are various ways of securing the mooring warps, depending on the circumstances. Use the Round Turn & Two Half Hitches to tie securely round an eyebolt or similar, and a Bowline to drop a loop over a large bollard. Ideally the end should be made fast ashore and the surplus warp hauled inboard then made fast at a suitable length. The warps can then be adjusted from the boat should it be required. This can be important in rough, exposed harbours with high walls when getting ashore is inconvenient or even dangerous. In a sheltered marina this problem does not arise so, although it looks most untidy, it is often more convenient to simply take the long end ashore, haul to the requisite length, then turn up round the cleat. With an eye or a bollard this is not feasible, and the short end should be taken ashore.

If a relatively small cleat is on the pontoon the eye of a Bowline should be doubled before putting it over, so as to reduce the movement and chafe. To double the loop simply twist the eye at the bottom so it forms an 8, then lay the bottom circle over the top one. Work the double eye round the cleat so that all parts are under even strain. If the cleat is open in the middle, you can pass the eye through the middle then loop it back over the outside to get a similar effect. If another bowline is already on the bollard you must dip the eye of the second up through it before putting it over. This will enable the first eye to be removed without having to lift off the second. See Photo 21 in Chapter 5.

If you lay alongside another yacht you must moor to him in just the same way as this, but you should also put extra head and stern ropes to the shore, so that the weight of your boat does not hang on the other's warps. Make sure all mooring warps at all times lead clear of obstructions and sharp edges that could chafe them or put an unfair strain

Photo 39 Note the thick plastic pipe threaded on this mooring warp where it passes through the fairlead. This takes the chafe and prevents the rope wearing away as it rubs in the fairlead.

on them, If this is unavoidable wrap some anti-chafe material round them, such as a short length of plastic hose or some old sailcloth. See Photo 39.

Approaching the Berth

As the yacht approaches the berth, stand clear of the helmsman's view - either ahead of the mast or on the side away from the shore. Be ready to step ashore from a point just ahead of the mast as this will be closer to the quay than the bow as the boat swings alongside. Do not jump if the boat is still moving towards the berth; wait until she arrives then step calmly ashore. The less you rush the less chance there is of you falling over or getting something wrong. Keep hands and feet clear of any part of the boat that might bang into a wall - the gunwale, for example.

Make sure the warp leads the correct side of rigging before stepping ashore, and keep a good eye open for the possibility of needing the wandering fender. If you see the need arising get the fender in place ready, and keep moving it as needed to ensure that it will actually be between yacht and jetty when the crunch comes. Then tie it rapidly to the guard-rail and nip ashore with your mooring warp before the boat bounces too far away from the berth.

It is worth practising the art of throwing lines as the ability to get one ashore or to another boat quickly and cleanly could save getting into trouble. This is best done by splitting the coil in two and taking one half in each hand. The half in the throwing hand wants to be as small as possible so that it will fly well through the air. Throw this half and let the rest run out freely from the other hand. If your mooring warps are rigged as shown in Figure 50 before you approach a berth, you will then be able to secure them instantly and properly the moment the shore end is made fast, and they will be able to safely take strain immediately as they will be in position in their fairleads. Always have the inboard end secured to a cleat to prevent losing it when the other end is hauled on.

Be prepared at all times in confined spaces for the skipper to suddenly order you to let go the anchor. In an emergency - engine breakdown in harbour, or if he misses a mooring and is in danger of hitting another yacht - anchoring promptly may keep you out of trouble. The skipper should tell you to have the anchor ready (see later section on anchoring), so you can let go immediately. Get the anchor away as quickly as you can, but do not panic or you will foul it up. Let out just enough warp to hold the yacht and no more, so that there is not slack to enable her to carry on and collide with the very boat you are trying to avoid. When the yacht is stopped, and the immediate danger averted, you can tell the skipper how much cable has been veered and ask for further instructions.

Mooring to a Buoy

There are two types of mooring buoy that you are likely to encounter. The commonest type has a large metal ring on the top to which you can tie the boat direct. The other, with no metal ring, is joined by a light line to a chain or heavy rope on the seabed to which the boat must be moored. The buoy has to be lifted on board and the light line hauled up to bring the chain inboard. The mooring chain or warp can then be secured to the boat, round the main anchoring cleat or *SAMSON POST* to provide maximum strength.

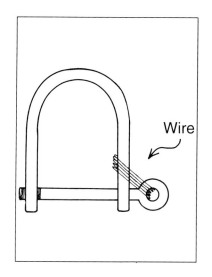

Wire

Fig 51 A shackle pin safely moused with wire so it cannot become undone. Special stiff wire called seizing wire is used for this, the ends simply being twisted together to secure it. Many other shackles on the boat should be seized like this, normally those out of sight up the mast holding the rigging together. Bend the twisted wire flat so that it cannot tear sails etc.

Mooring to the first type of buoy involves a slightly different routine as no buoy and *RISER* (line between buoy and seabed) have to be hauled aboard. The usual routine is to rig a *SLIP ROPE*. This is a strong, light line secured for'ard that is passed through the ring of the buoy and the end brought back aboard and secured so that the loop of line holds the boat temporarily to the ring of the buoy. A stronger warp can then be fastened direct to the ring with an eye-splice and shackle or a Round Turn & Two Half Hitches; or a length of chain shackled to the ring. The last is best as it cannot chafe either on the ring or in the bow fairlead. The shackle must be moused with wire as shown in Figure 51 so that the pin cannot work itself undone. See Photo 40. The most secure system is to pass the chain through the ring then shackle it back onto itself, as this avoids all risk of the pin being undone by constant motion in rough seas.

As the yacht approaches the buoy you should stand for'ard with the slip rope rigged ready, and indicate with your arm the direction of the buoy. See Photo 41. Have a boathook standing by in case you cannot reach the buoy with your arm; you can then hook and draw it towards you till you can reach with the slip rope. The skipper will probably also agree with you the signals to use for

Photo 40 Mooring lines should be passed through the bow fairlead, so if the anchor is stowed in the fairlead like this it will have to be lifted clear and brought onto the deck over the top of the pulpit. Make sure it is ready for letting go in case the skipper misses the buoy and there is no room to sail clear of other yachts.

Photo 41 The crew here is indicating to the skipper the angle of the anchor chain and the amount of chain that has been let out. The same signals can be used when approaching a mooring to show the direction of the buoy and the distance it is away.

128

indicating the distance off it - probably a number of fingers held up to show the range in metres. You will probably have to lie down at the last minute so that you can pass the slip rope through the ring easily, back on deck and cleated as quickly as possible. Make sure it leads under guard-rail and pulpit, and if possible through a fairlead.

The slip rope is an extremely useful device when mooring and letting go moorings, whether to a buoy or a jetty. It enables a temporary mooring to be made while the main one is cast off and tidied away, then it can be let go quickly and easily from onboard when you want to leave. It is, of course, imperative that the rope will slide reliably through the ring or round the bollard to which it is secured. Think how the rope will behave on being slipped, and ensure that it cannot catch on anything. Pull the loose end through slowly so it does not flick around and so jam between all and sundry. If it is rigged through a ringbolt on a jetty then let go the end passing down through the ring, so that when the other end is pulled from under the ring it tends to lift up instead of pressing down on and jamming the ring onto the loose end. If the ring is in the side of the quay then pull on the end leading to the outside of the ring.

Anchoring

In principle, there is no real difference between anchoring the yacht and anchoring the dinghy (see Chapter 5). The object of the exercise is to get the anchor warp laid neatly along the bottom as in Figure 24, so that the anchor is well dug in and the maximum amount of chain is providing friction against the sea-bed.

To achieve this the skipper calculates (either from the chart or by eye) where he must let go the anchor in order for the yacht to end up where he wants her. He then sails or motors into the tide and brings the boat to a stop in the required place. The anchor is then let go and as the boat drifts back with the tide the warp is veered steadily so that it lies out along the bottom. The skipper may come into the wind to do this if the wind is stronger than the tide, and he may go astern on the engine to dig the anchor in; he may even for various reasons let go the anchor while moving down wind or tide, but by and large your routine will be much the same.

Your routine consists of dropping the anchor into the water, then controlling the warp as it runs out so that it does not fall in a heap on top of the anchor. At the same time you should indicate to the skipper how the warp is 'growing' - ie the direction it leads in and the angle at which it goes down. This helps him to gauge how it is all lying on the bottom. Do this quite simply by pointing an arm in the required direction, and down at the angle of the cable.

How this anchoring process is implemented will depend on the yacht and her equipment. In a small boat you will lead the warp under the guard-rail or pulpit, as described earlier for preparing mooring warps, then simply pick up the anchor and drop it over the rail when ordered to let go. You then control the run of the cable by carefully pressing your foot on it, varying the pressure in order to vary the speed of VEERING. Hold yourself firmly in position while doing this - hang onto the forestay, perhaps - and never do it in bare feet or loose footwear. A larger boat may have a WINDLASS - manual or electric - which will enable you to simply loosen off the brake to let go the anchor (which will probably hang in the stem roller), then use the brake to control the cable.

To ensure that the cable runs out without jamming, a skipper will insist on the cable being FLAKED on deck before letting go, or flaked in the locker on hauling in. Flaking is a way of spreading out the chain or warp so that it runs freely. On deck it is laid fore and aft in lengths, each length being tucked close alongside the next. There is thus no bit of chain jammed under any other, which ensures that it runs without tangling. In the chain locker the cable may be flaked as it is hauled in and then run out direct from the locker without flaking on deck. If this system is employed it is good policy to flake on deck enough cable to reach the seabed, as the initial rush of the chain slows down when the anchor hits bottom, so a snarl-up is less serious. Chain should be flaked in a locker as on deck, but in layers flaked alternately fore-and-aft and ATHWARTSHIPS (across the boat).

Whatever the system for anchoring there should also be a system of marking the cable so that you can tell how much has been veered. This system may be blobs of paint on chain, lashings of small line on rope, or whatever. The skipper should tell you what his system is and you must watch carefully as the cable runs out so that you can keep track of it. The skipper will doubtless tell you how much cable he wants out and leave you to brake the cable on reaching this *SCOPE*.

Having laid out anchor and warp you must secure the warp to the strongest post or cleat for'ard. The *TUGBOAT HITCH* is ideal for securing particularly chain to a samson post, as it is extremely secure and cannot jam. It can also be tied after a couple of turns have been taken to hold the yacht, without having to remove the turns. Never make fast with a Clove Hitch as many do; if any weight comes on it you will never undo it. See Figure 52. Having secured the warp to the boat you should then check that the anchor is secured to the seabed. Hold the chain between the stem and the water and feel the vibrations. If the anchor is dragging there will be heavy, intermittent rumbling as it drags a bit then holds a bit then drags again. Tell the skipper immediately.

If there are old chains or similar rubbish on the bottom the skipper may secure a *TRIPPING LINE* to the anchor, so that it can be easily recovered if it gets jammed. This is simply a line long enough for the maximum depth of water with a buoy on the end. The other end is secured to the *CROWN* of the anchor (the end opposite the warp) and the whole lot thrown over the side with the anchor. Make sure it goes clear and does not tangle in the chain. If the anchor jams on the bottom it can be pulled out from the obstruction backwards by heaving on the tripping line.

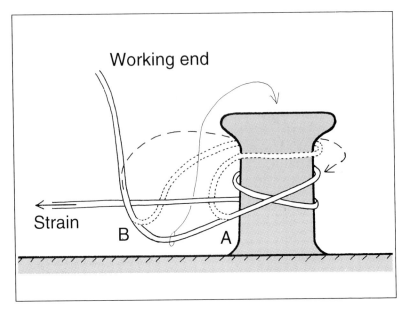

Fig 52 After taking a couple of turns on the bollard or samson post, tuck the bight under the standing part and loop it over the bollard, as shown by the continuous arrow and the dotted rope. Then take the working end round the bollard as shown by the pecked arrow. After this the bight can be tucked again, the whole process being repeated as often as necessary to hold the strain; slippery rope will need many tucks more then will chain for example.

Kedging

A *KEDGE* anchor is a smaller anchor than the main one and is intended for manoeuvring the yacht rather than holding her anchored. If a yacht runs aground, for example, the normal routine for getting her off (if you cannot simply motor or sail off) is to take the kedge anchor in the dinghy and drop it far away in the deep water. The crew can then heave on the kedge warp and hopefully pull the boat off and afloat again. Speed is important when taking out a kedge for this purpose, especially if the tide is ebbing, so an efficient routine should be worked out and used.

Photo 42 A kedge warp should be flaked in a figure of eight like this, then you can be reasonably sure it will run out without tangling. Note the wooden floorboards in this rubber dinghy, and the captive rowlocks that are permanently attached so that careless people cannot drop them over the sides.

The dinghy should be brought alongside the yacht for'ard and the kedge anchor lowered into it. The warp should then be flaked carefully into the dinghy so that it will run out without fouling - see Photo 42. As you row the dinghy furiously towards deep water the warp should be allowed to pay out astern. When you reach the end, drop the anchor over as quickly as possible before the tension in the warp pulls you back towards the boat, or the tide carries you back into shallow water. Make sure, however, that the warp is not tangled round the anchor. This system is much more efficient than simply dragging the warp out from the boat through the water, as rowing becomes increasingly difficult as ever more warp has to be dragged through the water.

Chapter 10
Life on Board

Essential though steering, sail handling and seamanship are
when you go cruising, they are not the full story. Domestic
life on a boat is quite different from that ashore, and it is a
great deal more important. Lives can be lost through
inattention to even the most apparently trivial of domestic
tasks, so it is vital that this aspect of sea-going life is both
thoroughly understood and carefully carried out.

Tidiness and Cleanliness

It is tempting when at home to think that an insistence on
these is simply a hang-up on the part of parents who have
nothing better to do with their time. Ships at sea, however,
have undoubtedly sunk in the past due to loose clothing and
suchlike being sucked into bilge pumps and jamming them.
The same has probably been caused by men reaching out in
pitch darkness during an emergency for urgently needed
tools and failing to find them because they have not been
returned to their proper stowages after use. And it is not
hard to imagine the trouble ensuing from the seasickness
and lethargy that are invariably caused by dirt and squalor
down below in small boats. See Figure 53.

Tidiness and cleanliness are not just important at sea,
they are vital. When the skipper shows you where to stow
your gear, stow it there immediately and make sure none of
it can come adrift. All the time you are at sea you should
stow every item of your gear away properly whenever you are
not actually using or wearing it, preferably in a plastic bag
or similar to keep it dry. In particular, do not dump gear
such as cameras, towels etc on the chart table as this can
be irritating in the extreme to the navigator; I have seen
stuff thrown over the side in such circumstances by irate
navigators trying to navigate. The same applies to ship's
equipment - tools, life-jackets and so on. *Always* return
them immediately after use to precisely the right stowage
place, then they can be found in the dark in emergency.

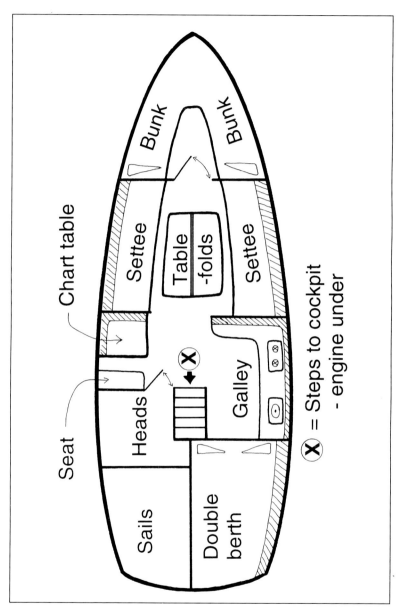

Fig 53 The interior layout of a fairly typical small yacht. Obviously boats vary, but this will give you a general idea of what to expect.

Keep yourself and the boat clean at all times. If you use a mug or plate, wash it, dry it and stow it away immediately after use. If crumbs are left in the galley after making sandwiches, clean them up right away. Keep the gash bucket covered and do not let it overflow. The skipper will doubtless have a system for dealing with the rubbish; make sure you know what it is and adhere to it. Squalor accelerates very rapidly on a boat due to the confined space and constant movement, and the plummeting disintegration of morale that it can cause is hard to believe unless you have experienced it. Learn from others' unfortunate experiences rather than trying it for yourself.

Using the Heads

HEADS is a nautical expression meaning lavatory. It stems from the old days when it consisted of a piece of wood with a hole in that jutted out over the side in the head (bow) of the ship. Some people use the word in the singular. Singular or plural it is a more complex business than relieving yourself ashore, and failure to operate the device correctly can result in the skipper having to dismantle the whole thing to clear blockages. This is unlikely to make you popular on board, so it is as well you learn how to use it properly.

There are many different types of heads available these days but the basic difference between all of them and a shore-side lavatory is that the waste matter must be pumped out of the bowl and the flushing water pumped in. Sometimes two pumps are used - one for each job - and sometimes one does both; sometimes they are electric; sometimes they pump into the sea and sometimes they pump into a tank on board, which is then discharged out at sea later rather than close inshore or in a marina. Whatever the precise system there should be clear instructions posted nearby: read them and follow them to the letter.

Relieving yourself over the side of the boat (men and boys only!) can be a dangerous business at sea due to the risk of being jolted overboard while not hanging on securely. Some skippers will not permit it; if yours does, the safest way is to lean against the lee SHROUDS (rigging) with an arm securely round each one.

Water and Electricity

Both these commodities are usually in very short supply on a small boat, so they must be conserved. Water may have to be carried long distances in cans when you are cruising, so running short can cause major problems. Electricity is stored in batteries in relatively small amounts; when it is used up the engine has to be run for long periods to recharge these batteries, and this can be most unpleasant and noisy when you are relaxing in a peaceful anchorage. Never let water run from a tap and never leave on lights or other electrical equipment unless absolutely necessary.

A mugful of fresh water should be enough to clean your teeth then have a basic wash with a flannel. While avoiding an accumulation of dishes in the galley sink, try to wash a reasonable load at once rather than constantly doing odds and ends. See Photo 43. Many boats have a pump in the galley that draws salt water direct from the sea. Skippers may insist on washing up, vegetable peeling and so on being done with this; perhaps also all-over washing of the crew, followed by just a rinse in fresh water. Special soaps, shampoos and detergents can be had that will lather in sea-water, and these will doubtless be provided by such skippers.

Boat electricity is a complex subject and this is not the place for a detailed discussion of it. A general idea of its basic principles will, however, help you understand the importance of conserving it. It is the same low-powered electricity that is used in a car and is made in the same way - by a generator driven off the engine. It is then stored in a battery just as it is in a car, for use later as required. This battery is no different in principle to the one operating a cassette recorder or electronic toy, just very much bigger. As it is used (for lights, radios etc) so it runs down until finally it becomes completely empty of electricity. It then has to be filled up again (recharged) by running the engine, probably for quite a few hours. Normally this is done long before the battery is completely empty of electricity.

To complicate matters, in some installations serious problems can arise if the battery gets too low as the engine

Photo 43 Making the tea will undoubtedly be one of your jobs! Note the position of the fire blanket close to the doorway but near the cooker. It should never be positioned so that you have to reach over the cooker to get to it. The big white box next to it is a luxurious gas water heater. Note the sturdy post in the foreground for you to hold onto in rough seas.

generator may not then be able to charge it; the battery may have to be taken ashore and charged with special equipment. In a very badly designed installation it may even be impossible to start the engine when the battery has run down, although most boats will be equipped with a separate battery reserved purely for starting. Under no circumstances must this ever be used for any other purpose, although a changeover switch may well enable you to do so.
See Photo 44.

Photo 43a A small yacht galley with sink and neatly arranged stowage for food, crockery etc.

Photo 43b Part of the saloon of a small yacht, showing the first-aid kit, fire extinguisher, barometer, navigation instruments and radio for weather forecasts. See how they are stowed so as to take up as little space as possible, and also to be near to hand when needed.

Photo 44 A corner of the chart table showing an electronic navigator at bottom left and VHF radio bottom right. Above this is an ordinary radio for weather forecasts etc, then a switch and fuse panel for the boat's electrical system.

It is very easy to leave lights on undetected; to forget the anchor light in the morning so that it burns all day; to wander off and leave a radio blaring, and so on. You must be constantly alert to these things and aware of the enormous trouble that might be caused if the battery is allowed to run down too much, due to excessive use of electricity.

Gas and Bilge Water

These are both potentially serious dangers on a boat. Cooking gas is heavier than air, so if it leaks it falls down into the bilge and stays there. It does not blow away in the wind whenever the door is opened as it will in a house, so it steadily accumulates and eventually an explosive mixture is formed with the air. This can be ignited by the minutest spark - even that caused by a light switch being turned on - so it is vital to guard against it. There are two lines of defence: preventing the gas from leaking, and removing it from the bilge before it can make up an explosive mixture. Both are done quite simply by religiously adhering to these golden rules:

1   Always turn off gas at the bottle or isolating tap when not using it. This reduces considerably the number of joints from which it could leak.

2   Strike the match first, then open the tap and light the gas immediately before any can pour out into the bilge.

3   If you smell gas, turn off the bottle and tell the skipper - immediately.

4   When bilges are pumped out continue with a dozen or so dry strokes to pump out any gas that has leaked.

The danger from bilge water is perhaps more obvious and certainly more visible. Quite simply, if there is too much of it the boat will sink. There are various ways water can get into the boat and fill her up while you are all ashore enjoying yourselves, but they are all very simply guarded against. The main danger is of pipes falling off *SEACOCKS* (valves permitting the ingress of water through the hull) and this is guarded against by always keeping them shut when they are not actually being used. This includes heads inlet and outlet cocks, galley sea-water inlet cock, engine cooling water inlet cock and so on. The skipper should tell you about these, but it is important to appreciate the possible dangers, even when an installation is brand new and in apparently perfect condition. There are a great many things that could cause a pipe to come adrift from a seacock long before the fault can be seen by anyone - electrolytic corrosion of the seacock pipe, loose pipe clips, vibration from the engine, a kick from someone's boot etc, etc. If the cock is closed, no harm is done; if it is open, the boat could sink - it is as simple as that. Keep them closed. A bonus is that the constant opening and closing prevents them from seizing up.

General Hints

Life at sea tends to harbour rather more extremes than that ashore, and two that can cause an awful lot of misery are the heat of the sun and the cold of the wind. You will find the sun has greater burning power on the water than the land, even though it may not feel especially hot, so strong sun-barriers need to be used, particularly on sensitive skin like the nose. It helps to wear a floppy sun-hat to keep the sun off your head and provide some shade for the face. Salt water on the skin seems to make the sun worse.

The wind is quite the opposite, being much colder on the water than you expect. Most of the time, certainly when day-sailing, you will not need masses of clothes, but you should have something wind-proof like an anorak or smock, and a jumper, however hot the day. If the wind freshens, or you find yourself in the shade, or night begins to fall, it will very quickly become cold; a most uncomfortable few hours could then ensue if you have a long sail home with no warm clothes.

Seasickness can strike on even a relatively calm day, and be most unpleasant. If you feel queasy, tell the skipper immediately. He will probably make you eat something dry to settle your stomach, and give you work on deck (steering perhaps) to take your mind off it. The more you can do to avoid thinking about it the better you will feel, so do not curl up miserably in the corner of the cockpit - keep busy. Do not go below for even a moment, as this may easily make you instantly sick. If you suffer badly you must sort out some treatment with the skipper; it is a most debilitating thing to put up with. Most people do, however, get over it after the first couple of sailing trips.

Safety Routines

There are certain dangers involved in even the normal daily routines on a boat and it is vitally important to be aware of them. These dangers are: fire from fuel leaks, cooking accidents etc; suffocation from lack of ventilation; making water from faulty sea-water pipework etc; injuries and medical troubles far from shore.

A good skipper will have calculated routines laid down both for minimising these risks and for dealing with the emergencies if they do occur. Specific emergency routines were discussed in Chapter 1, so let us now look at the general precautionary routines that should be implemented during daily domestic life on board.

We have already discussed the dangers of leaking gas and leaking water, so let us consider the others. Fire is a very serious risk on a boat as although there is plenty of water about it is very difficult to stand back from the fire and use it. A boat is so small that you will be virtually IN the fire wherever it is, and of course water is not always the best thing for putting out fires. Prevention of fire is therefore of vital importance, and the greatest care must be taken with anything that could cause it - cooking, burning off paint, refuelling and stowing petrol, smoking, lighting oil lamps and so on. This is directly the concern of the skipper, and he should instruct you on precautions to be taken.

People are not infrequently suffocated on boats when they operate cookers or heaters to keep warm with all the hatches etc closed. As heaters burn so they use up the available oxygen; if there is not sufficient ventilation to replace this oxygen it will become steadily depleted in the cabin until eventually there is not enough to sustain life. The problem is that the effect is not obvious - people do not suddenly feel ill and have to open the hatches, they gradually get pleasantly sleepy, then doze off, then die. *Never* use cookers or heaters of any sort without ensuring that hatches, portholes or ventilators are open.

Medical emergencies are, of course, the responsibility of the skipper, but you should familiarise yourself with both the whereabouts and the contents of the first-aid kit, as well as a first-aid book. You should appreciate that even only a few miles offshore you could be a very, very long way from a doctor or hospital, so the greatest care must be taken to prevent accidents and illness.

You should consider the possibility that all those senior to you on board may be ill or injured, with the boat a long way from help. You may be the only person available to treat them, so at least be prepared mentally to read the book and do what it says. Help can be summoned by various means and you should study them carefully in Appendix 3. See also the section on fixing your position in Chapter 8.

Loss of the Skipper

This is a potential disaster that few yachtsmen seem to consider. In a large, traditional sailing boat the command would be automatically assumed by the First Mate, and if anything happened to him too, the Second Mate, and so on. A small yacht rarely carries a crew with such a useful, clearly-defined hierarchy of capable hands, so the loss of a skipper could leave a young, inexperienced crew adrift, far from safety and with probably very little, if any idea what to do.

It is highly likely, of course, that trying to do too much will prove even worse than doing nothing. The concept of first-aid - doing just enough to prevent the situation worsening while waiting for help - should be applied here also. What is needed is a very simple list of actions that any reasonably capable crew can carry out, that will maintain the safety of the yacht while at the same time summoning rescuers to an accurate position. In Chapter 8 we discussed how to take a position from a radio navigator and plot it on the chart, and also how to update this position if the navigator fails. In Appendix 3 is explained the simple details of how to make a distress call on a VHF radio-telephone. A proper distress call containing an accurate position will bring help to you in very short order indeed, so it is most important that you learn this procedure thoroughly.

If you have no working radio-telephone you must call for help with the red flares, but it must be done methodically if you are to create the best chance of success. Parachute flares should be used initially to alert others to your predicament, as these can be seen a very long way off. If you are near the coast, someone ashore may see them and phone the coastguard, so you are not totally dependent on passing ships. It is best to fire them in pairs, the second perhaps two minutes after the first, in order to be spotted by someone who catches just a glimpse of the first then stares in that direction for a while. Flares should be fired downwind and at 45 degrees to the vertical as this causes them to reach maximum height.

When a vessel appears reasonably close you should deploy the red hand flares, holding them on the leeward side of the boat, so that a rescuer has a continuous red light to home on. Use them sparingly, in case the first ship does not see you. Wait till he gets as close as possible before firing them off.

Keeping the yacht and her crew out of danger while waiting for assistance may not be easy, depending on conditions. As a simple rule, however, you will be unlikely to go far wrong if you heave-to under sail (see Chapter 7) or set the rudder to make the boat run in circles if under power. In very rough seas, however, this could be dangerous; it will be better to motor as slowly as you possibly can into the waves. When heaving-to, do so on the tack that makes the boat sail slowly away from danger - shore, sandbanks etc. You must learn how to operate the engine controls so that you can put in and out of gear and adjust the speed, and also stop and start it. Having given a position to rescuers it is important that you remain as close to it as you can.

Put life-jackets on all the crew and prepare the liferaft for launching, but do not be tempted to abandon ship in a panic - the yacht will almost certainly be a safer place than the liferaft (and much easier for a rescuer to find), so stay aboard, even if you have to pump continuously to keep afloat. There is an old saying that you should only ever step UP into a liferaft - in other words, if the yacht sinks under you. If you do have to abandon ship, take the flares.

Photo 45 It is not all hard work and discipline. These youngsters enjoy a relaxing sail in sheltered coastal waters while the adults do the work. Just below the pulpit you can see the furling drum for the roller jib.

Appendix 1

Recording the Shipping Forecast

This is a job that may well be given to you and it is an extremely important one. It is not easy, as the forecast is often read at great speed, but it is vital that you record it accurately. Some form of shorthand together with a routine technique will make this very much easier. This is helped by the fact that the information is always read out in the same order - Time of issue; gale warnings in force; general synopsis; area forecasts; station reports. Area forecasts consist of: areas covered; wind direction and strength; weather; visibility. Station reports consist of: station; wind; weather; visibility; barometer reading and movement.

As you know what each part of the forecast refers to, you will be able to use very concise shorthand to save time. Areas having the same forecast can be grouped together and their forecasts abbreviated; eg **HTDWPP NW3/4 b SW5–6l sh-rl G-M** for Humber, Thames, Dover, Wight, Portland, Plymouth: north-westerly 3 to 4, backing south-westerly 5, increasing 6 later; showers, rain later; good becoming moderate.

The use of preprinted forms such as Metmaps make the recording of the forecast even quicker, as you can see from Figure 54. The often complex detail in the General Synopsis is much quicker and easier to note on the picture than to write down, although some geographical knowledge will be required to know the places referred to. Some understanding of the terminology will also help. A Low is an area of low pressure, generally producing wet and windy weather, while a High is high pressure usually giving light winds and sunny weather. This is a gross over-simplification but it will help to give you a general idea of the significance of the information you are noting.

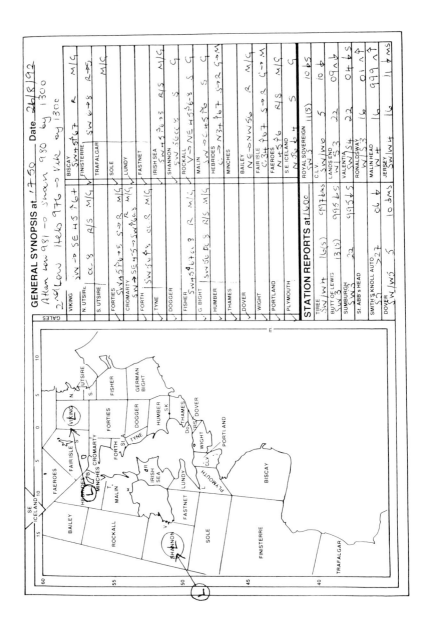

⇐ **Fig 54**

This shows a fairly typical Shipping Forecast noted down accurately but quickly and concisely. Go through the forecast carefully and see if you can work out what is happening. The general Synopsis gives an Atlantic Low moving east into Shannon, and another Low moving from Hebrides to Viking. You can see how these two are marked on the chart of the areas. The Station Reports give present weather conditions at various places around the coast; these can be very useful for assessing the rate of movement of the weather patterns.

The barometer is an instrument that measures atmospheric pressure, so the information it gives on board helps the skipper enormously in deciding where the yacht is in relation to the weather systems. The barometric pressure and its movement (rising or falling) given in the Station Reports helps him assess where the high and low pressure areas are geographically, and the sum of all this information enables him to forecast the weather the yacht is likely to experience for some hours ahead. The more often you listen to and record the Shipping Forecast and discuss it with the skipper, the more you will learn about this vitally important subject.

Weather Information

Gale Warnings

Imminent	- within 6 hours of issue by Met. Office
Soon	- 6 to 12 hours after issue
Later	- over 12 hours after issue

Visibility in Sea Areas

Good	- over 5 nautical miles
Moderate	- 2 to 5 nautical miles
Poor	- $^1/_2$ to 2 nautical miles
Fog	- less than $^1/_2$ nautical mile

Speed of Pressure Systems in Forecast

Slowly	- less than 15 knots
Steadily	- 15 to 25 knots
Rather quickly	- 25 to 35 knots
Rapidly	- 35 to 45 knots
Very rapidly	- over 45 knots

Barometer Changes (rough guide only)

Over 5 Mb in 3 hrs - almost certain Force 6
Over 8 Mb in 3 hrs - almost certain Force 8

The Beaufort Wind Scale

FORCE	KNOTS	WAVES(ft)	LIKELY SEA STATE IN OPEN WATER
0	0	0	Flat calm. Any swell is not caused by wind
1	2	0	Patches of ripple on surface
2	5	1	Ripples all over surface
3	5-10	2-3	Occasional white horses on wavecrests
4	10-15	4-5	Many white horses on wavecrests
5	15-20	6-8	Waves cresting, with spray blown from them
6	20-25	8-12	Streaks of spray and foaming crests
7	30-35	12-16	White foaming crests, whipped away in gusts
8	35-40	20-25	Rough and disturbed, with 'boiling' patches
9	40-45	25-30	Covered in white foam. Spray reduces visibility
10	50-55	30-40	Visibility badly affected by blown spray
11	60-65	45	Air full of spray, causing very poor visibility
12	65+	45+	Visibility almost zero in driving spray

Fig 55 This is convenient shorthand for noting wind strengths, 'force 6' being more concise than 'somewhere between 20 and 25 knots'. Wind strength is very erratic, often varying between for example 10 and 40 knots in the forecast for Forties in Figure 54, so there is little call for great precision.

Appendix 2

Keeping the Logbook

This is another important task that is likely to be delegated to you when you are on watch. The logbook constitutes a record of the passage, detailing all navigational information such as position, course and speed, features sighted, fixes taken and so on, together with regular recordings of the weather and barometer. The former information enables an accurate navigational plot to be maintained on the chart, while the latter enables the skipper to assess the likely future weather. The accurate and regular recording of all this information is extremely important; the safety of the yacht and her crew may depend on it if the weather turns foul or the visibility closes in off a dangerous coast.

Even a yacht equipped with electronic navigation aids, radar etc should keep this record in the logbook, so that navigation may be picked up from it and continued by traditional means if the electronics fail. As with recording the Shipping Forecast, a system of abbreviations is needed in order to fit all the information into the small space available. In Figure 56 you can see an example of a logbook page filled in, and Figure 57 shows the shorthand symbols that have been used. As the logbook constitutes a permanent record of the passage you should agree a set of symbols with the skipper, so that others can read and understand the entries at a later date.

There are various ways of keeping a logbook and your skipper may well have strong views on how his should be kept. The one shown here is an example from a logbook I designed specially for yachts, and it provides the flexibility to record information either regularly on the hour (best for long passages) or simply when convenient or useful (as here; perhaps best for coastal sailing).

Record intervals 1

Insert record every hour or so

Position 2

Record positions as lat/long or Range & Bearing from known point. Bearing is from point towards your position. Range is in nautical miles. Record all bearings in °M (see Fixed-Time Specimen)

Repeated entries 3

Blank spaces are less cluttered than " (ditto) marks but may mean entry has been forgotten! (See Fixed-Time Specimen)

Unsettled weather 4

Record details when in harbour in order to assess trends

Narrative 5

Little needed for a simple delivery trip

DATE FRIDAY 14 JUNE 92   FROM   Denhaven

TIME GMT	COURSE ORDERED °C	LOG READING	COURSE STEERED	DIST RUN	LEEWAY	WIND	SEA	WEATHER	Vis	Bar	POSITION	SOURCE OF FIX	NEXT WP
0740	⚓	-	-	-	-	NW3	Calm	0/8	G	15	Denhaven Hr	Vis	1
0805	Var	0	Var	0	-						WP1	RN	4
0830	240	1·2	240	1·2	nil					15½	Bench Bn ← 1'	Vis	
0855		2·5						slight			½' → Griddle Pier	Vis	
1000	250	6·5	250			NW4	mod				54°11'3N:01°13'7E	RN	
1115		12·0								15	WP4	RN	7
1230	185	18·1	185							14	190° Cresswell Bn 3'	RR	
1400		25·6						½ H		13	53°55'1N:01°07'4E	RN	
1510		30·2				WS				11	WP7	RN	8
1630	230	35·1	230			S5/6	M/R	⅜M	M	09	53°50'0N:01°02'6E	RN	
1740		39·4				S6/7		⅞LoR		04	WP8	RN	9
1910	180	47·6	180				red			99	Bell Pt DIP 210	CB	
2010		52·2					slight	⅞L-R		95	WP9	Vis	10
2030	⚓	54·3				S 6%				93	Oxley Marina	Vis	-
2330						SW8	-			81			

NARRATIVE

No problems
Will try to get round to Pintle Creek by Sunday

Advantages of FLEXI-TIME System

1. More than one row can be used per entry, allowing more space for Remarks when busy

2. Entries may be made for the time of passing waypoints, charted features etc.

Fig 56 This is a specimen page from a logbook, with some useful comments on the entries. This 'Flexi-time' system records entries when events occur - the taking of a fix, altering course, starting the engine and so on - endeavouring to do so at roughly hourly intervals. The 'Fixed-time' system referred to records full information every hour on the hour, so as to simplify navigation on long passages.

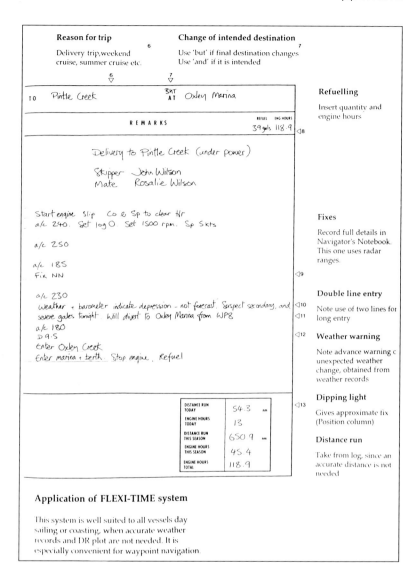

Reason for trip 6

Delivery trip, weekend cruise, summer cruise etc.

Change of intended destination 7

Use 'but' if final destination changes
Use 'and' if it is intended

TO Pintle Creek   3HT AT Oxley Marina

Refuelling

Insert quantity and engine hours

REMARKS   REFUEL 39 gals   ENG HOURS 118.9

Delivery to Pintle Creek (under power)

Skipper John Wilson
Mate   Rosalie Wilson

Start engine Slip   Co & Sp to clear H/r
a/c 240. Set log O. Set 1500 rpm. Sp 5kts

a/c 250

a/c 185
Fix NN

a/c 230
Weather + barometer indicate depression - not forecast. Suspect secondary, and
severe gales tonight. Will divert to Oxley Marina from WP8
a/c 180
D 9.5
Enter Oxley Creek
Enter marina + berth. Stop engine. Refuel

Fixes

Record full details in Navigator's Notebook. This one uses radar ranges.

Double line entry

Note use of two lines for long entry

Weather warning

Note advance warning of unexpected weather change, obtained from weather records

Dipping light

Gives approximate fix (Position column)

Distance run

Take from log, since an accurate distance is not needed

DISTANCE RUN TODAY	54.3	nm
ENGINE HOURS TODAY	13	
DISTANCE RUN THIS SEASON	650.9	nm
ENGINE HOURS THIS SEASON	45.4	
ENGINE HOURS TOTAL	118.9	

Application of FLEXI-TIME system

This system is well suited to all vessels day sailing or coasting, when accurate weather records and DR plot are not needed. It is especially convenient for waypoint navigation.

GENERAL	
⚓, Hr	in harbour (anchored, moored), harbour
Bn, By, Lt	beacon, buoy, lighthouse
WP, XTE	waypoint, cross-track error
Co, Sp, Kts	course, speed, knots
a/c, Var	alter course, various
←, →	left-hand edge of land, right-hand edge of land
↦, ↤, φ	abeam to starboard, abeam to port, in transit
DIP	light dipping over horizon (eg Bell Pt DIP 210)
L 42.7	log reading = 42.7 nautical miles
D 13.4	depth = 13.4 metres
Tack port 140	put about onto port tack; heading 140°(C)
2 rolls Mn	tuck 2 reefs into mainsail
Roll gen	roll up genoa
Set No 2	change to No 2 jib
NN	navigator's notebook
DR	dead reckoning
TPL	transferred position line

WEATHER	
C, S, M, R	calm, slight, moderate, rough sea states
R, Sh, Sl, Sn	rain, showers, sleet, snow
s, m, h, c	slight, moderate, heavy, continuous (rain etc)
G	good visibility (over 5 nautical miles)
M	moderate visibility (2 to 5 miles)
P	poor visibility (half to 2 miles)
F	fog (less than half a mile)
3/8	fraction of cloud cover
H, M, L	high, medium, low cloud

SOURCE OF FIX	
RN	radio navigator (Decca, Loran)
SN	satellite navigator
DF	DF bearings
CB	compass bearings
RR	radar ranges
EP	Estimated Position
Vis	approximate visual fix by eye

Fig 57 The sensible use of abbreviations such as these saves a huge amount of time and space, and also improves clarity. Use the present tense for remarks as it is usually shorter than the past.

Appendix 3

Radio Distress Procedure

Tune set to
Distress Frequency - VHF=Channel 16: MF=2182 kHz

Transmit following Distress Call on High Power:

MAYDAY MAYDAY MAYDAY this is (boat's name 3 times)

MAYDAY this is (boat's name)

Transmit Distress Message:
Position of boat (bearing & distance *from* landmark is best)
Nature of emergency (sinking; on fire; etc)
Type of assistance needed (pumps; fire fighting; etc)
Number of persons on board (so all can be accounted for)
Any other useful information (rate & direction of drift etc)

Listen for a reply.

Repeat complete call until reply heard.

For problems less than grave emergency substitute for the
word MAYDAY:
PAN-PAN = urgent message concerning safety of crew or
ship
PAN-PAN MEDICO = urgent request for medical help
SAYCURITAY = important message concerning navigation
or weather danger (report floating wreckage and suchlike)
   Keep the radio set tuned to Distress Frequency when
not in use. Make sure you know how your set operates.
Practise making distress call with radio off; try to use set
as much as possible so that you become familiar and
confident with it.

Distress Signals

1   Gun or explosive signal fired about once a minute.

2   Continuous sounding of foghorn.

3   Red flares - rocket, parachute or hand.

4   Orange smoke signal (hand flare or floating canister).

5   Flames on the vessel (rag on boathook soaked in paraffin).

6   **S-O-S** sent in Morse by any means (usually light or sound).

7   The word **MAYDAY** on a radio-telephone (see above).

8   International code flags - **N** above **C.**

9   A square flag and a round shape hoisted together.

10 Slow, repeated raising and lowering of outstretched arms.

11 Built-in alarm signals from radio transmitters.

12 Signals from Emergency Position Indicating Radio Beacons (EPIRBS).

13 Ensign hoisted upside down.

14 Ensign made fast, high in the rigging.

15 Article of clothing attached to oar - blowing horizontally

Use only in situation of grave danger to vessel or crew

Glossary of Nautical Terms

There are a great many specialised words associated with boats and the sea and you cannot hope to learn them all immediately. The ones printed in *ITALIC CAPITALS* and explained in the text are all you need for the time being, but the complete list here will enable you to browse through and learn others as time and your inclination permit.

Abaft	behind
Abeam	directly out to the side of the boat
Adrift	loose; late; broken off
Aft	at the back of the boat
Ahead	forward; in front of the boat
Amidships	in, or in line with middle of boat
Arm the lead	put tallow in lead-line lead to sample seabed
Astern	backwards; reverse; behind the boat
Athwartships	across the boat
Autopilot	automatic steering gear
Awash	level with the water
Aweigh	anchor is aweigh when just off seabed
Back	wind backs when it shifts anti-clockwise
Back Eddy	water flowing in the opposite direction to main flow
Ballast	weight low down in boat to give stability
Batten down	secure hatches etc firmly before going to sea
Barometer	instrument that shows air pressure
Beam	width of boat at widest part
Bear off	push off from jetty etc
Belay	secure rope round cleat; cancel order
Below	down inside the cabin
Beneaped	stuck aground with tide changing to neaps
Bight	loop in rope; middle part of rope
Bilge	inside of boat at very bottom
Bilge pump	removes water from the bilge
Binnacle	stand on which a compass is mounted
Bitter end	very end of a rope or chain cable

Boathook	pole with a hook on the end
Bollard	large post for mooring boats to
Bow	front end of a boat
Bowline	important knot for making an eye or a loop in a rope's end
Breast rope	short mooring line straight across to shore
Bring up	come to anchor
Broach	swing violently round broadside to waves
Bulkhead	internal partition wall in boat
Buoy	sea mark anchored as guide to navigation or special ones for mooring
Buoyancy aid	foam-filled waistcoat that helps you float
Cast off	let go of a line or mooring
Centreboard	a vertical plate that is lowered through the bottom of dinghies to prevent leeway
Chart	a map of the sea showing depths
Choke	an engine control that helps starting outboard motors
Cleat	T-shaped fitting to which ropes can be secured
Clew	lower aft corner of a sail
Coaming	raised surround to cockpit etc
Cockpit	well at stern (or amidships) for crew to sit in
Compass	always points North, so vital for navigation and steering
Courtesy ensign	small ensign of country being visited
Crown	bottom end of anchor, where flukes begin
Dan buoy	thrown after 'man overboard' to locate his position
Deck	flat surface on top of a boat
Deckhead	underneath the deck (inside the cabin)
Dinghy	small open boat, often used as tender for yacht
Drag	anchors drag when they are not holding
Draught	depth of boat below waterline
Ebb tide	flows out of harbour and falls in height
Fairlead	smooth fitting at deck edge to lead ropes ashore or to lead jib sheets
Fair tide	tide that is running with you

Fairway	clear, navigable channel
Fall	loose length of rope leading away from a tackle
Feather	to angle oars to reduce windage
Fender	soft plastic sausage to cushion boat against wall
Fetch	distance across water in which waves can build up
Fiddle	rail round cooker etc to stop things sliding off
Flake	lay rope or chain in loose figure of eight
Flare	emergency fireworks to attract attention in distress
Flukes	points of anchor that dig in seabed
Flood tide	tide flowing into harbour and rising in height
Fo'c'sle	compartment right forward in boat
For'ard	commonly used contraction of forward
Foul anchor	anchor tangled up in chain or rough seabed
Foul tide	tide running against you
Freeboard	height of deck above waterline
Galley	kitchen onboard a boat
Gimbals	pivots enabling compass etc to stay upright
Goose-wing	sailing downwind with mainsail on opposite side to headsail
Ground tackle	generic term for anchors and warps etc
Guard-rail	wire fence around deck edge
Gunwale	corner between topside and deck (pron: gunnel)
Gybe	to turn a boat sailing downwind so that the wind blows over the other quarter
Hand	to lower a sail
Hand bearing -compass	small portable compass for taking bearings of ships and land
Hank	a clip that attaches the luff of the headsail to the forestay
Head	the top corner of a sail
Head rope	mooring line from bow to jetty

Headfoil   A grooved foil on the forestay into which the luff of a headsail may be inserted

Heads   boat's WC

Heave-to   To stop a boat's forward movement by backing the jib and adjusting the helm

Heaving line light line with heavy knot on end to throw ashore

Helm   steering position; tiller or wheel

Helmsman   person steering the boat

Inboard   in or on the boat (inboard engine etc)

Jury   makeshift (rudder, rig etc)

Keel   backbone of boat at the bottom; weight under boat

Kedge   smaller anchor for temporary use

Knot   speed of one nautical mile per hour

Lead-line   line with lead weight on end to check depth

Lee shore   shore towards which the wind is blowing

Lee side   side of boat opposite the wind

Leeward   away from the wind

Leeway   sideways drift of boat caused by the wind

Lifebuoy   Buoyant horse-shoe shaped emergency device to aid persons in the water

Lifejacket   device that will keep unconscious person afloat

Liferaft   usually an inflatable last-ditch resort for abandoning ship

Log   device to measure distance run through water

Logbook   to keep detailed records of a passage

Lubber's line fixed line on compass showing boat's heading

Luff   the leading edge of a sail

Make fast   secure a line to a cleat etc

Marlinspike   tapered steel bar for undoing shackles etc

Millibar   unit of measurement for atmospheric pressure

Mooring   buoy anchored to seabed for tying up to

Mouse   secure pin to shackle with wire

Nautical Mile about 2000 yards; one minute of latitude

Neap tide	occurs two weeks after Spring Tides with smallest tidal range
Offing	area of sea away from the shore
Outboard	outside the boat (outboard engine)
Outhaul	pulls mainsail clew to end of boom
Painter	mooring line permanently attached to dinghy
Port	left hand side of boat looking forward
Preventer	line from main-boom led forward to prevent accidental gybe
Pulpit	strong guard-rail round bow
Pushpit	strong guard-rail round stern
Quarter	after corner of a boat
Range	difference between tidal height of High Water and Low Water
Reef	to reduce sail area, usually on mainsail
Reeve	to pass a rope through a block
Riding light	another name for an anchor light
Ringbolt	bolt with large ring attached, for mooring up to
Riser	cable from seabed mooring to buoy on surface
Rowlock	U-shaped gap in gunwale to rest oar in
Rubbing strake	half-round timber or moulding to protect gunwale
Rudder	flat plate swivelling at stern to steer boat
Safety harness	worn by crew and attached to the boat to prevent falling overboard
Samson post	heavy post on foredeck for anchoring or mooring
Scope	amount of anchor warp in use
Seacock	a through-hull valve that can be closed to prevent flooding
Sea-legs	you have your sea-legs when used to motion of boat
Seizing	secure lashing holding two ropes together
Shackle	U-shaped metal fitting closed by screwed pin
Shank	long arm of an anchor
Sheer	swing about at anchor; curve of boat's deckline

Sheet	primary control of a sail
Shoal water	shallow water
Shrouds	wires that support the mast at its sides
Slack water	period of no tidal stream at High and Low Waters
Sounding	depth of water on chart; depth measured by lead
Spinnaker	large nylon downwind sail set flying and attached to a pole
Spring	mooring warp that prevents movement along jetty
Spring tide	those with the largest range between High and Low Water
Stanchion	post supporting guard-rail
Standing part	end of a rope that is secured to something
Starboard	right hand side of boat looking forward
Steerage way	just sufficient speed for the rudder to steer
Stem	the very front edge of the bow
Stern	the back of a boat
Stern rope	mooring line from stern to jetty
Stock	cross piece on anchor making it fall over and dig
Surge	ease warp gradually round cleat when under strain
Tackle	rope rove through blocks to increase power
Tender	dinghy for getting ashore from boat when anchored
Thwart	athwartship seat in small boat
Tide	vertical movement of the sea caused by the sun and moon's gravitational effect
Tidal stream	the current created by tidal movements
Tiller	long stick to turn rudder when steering
Topping lift	running rig to support the boom when the mainsail cannot do so
Topsides	side of boat above waterline
Transom	a flat vertical stern
Trim	fore-and-aft attitude of boat in water
Under way	not attached to the land (not necessarily moving)

Up and down	when anchor cable is vertical
Veer	wind veers clockwise; to let out cable or warp
Wake	trail of disturbed water left behind moving boat
Warp	general term for mooring ropes, anchor cable etc
Watch	spell of duty for crew - steering etc
Weather shore	shore which the wind is blowing away from
Weather side	side of a boat facing the wind
White horses	waves breaking with white foam
Windage	area of boat above water that wind blows against
Windlass	a powerful winch on the foredeck used for the anchor cable
Wind-rode	lying to the wind when moored or anchored
Windward	towards the wind
Working end	free end of rope, used to tie knots in
Yaw	swing from side to side of course being steered

Index